CURIOUS NATURALISTS

NIKO TINBERGEN, a native of the Netherlands, taught at the University of Leiden, at Yale and Columbia, and since World War II has been at Oxford University as university lecturer in animal behavior. He has long worked closely with the noted student of animal behavior, Konrad Lorenz. Dr. Tinbergen is the author of *The Herring Gull's World*, *The Study of Instinct*, and a number of other writings in this field.

CURIOUS NATURALISTS

NIKO TINBERGEN

PUBLISHED IN CO-OPERATION WITH
THE AMERICAN MUSEUM OF NATURAL HISTORY

THE NATURAL HISTORY LIBRARY
ANCHOR BOOKS
DOUBLEDAY & COMPANY, INC.
GARDEN CITY, NEW YORK

Curious Naturalists was originally published in this country by Basic Books, Inc. in 1958. The Natural History Library edition is published by arrangement with Basic Books, Inc.

Natural History Library edition: 1968

American Museum of Natural History
Members' Edition: 1969

PREFACE

This book describes the activities and some of the discoveries of a small number of naturalists who have joined me, at one time or another, in the pursuit of our common hobby: the study of the behaviour of animals in their natural environment. It covers about 25 years of biological sightseeing and exploration and tries to tell, without the use of technical language, a story of our field work and of the joys of discovery.

To publish or not to publish has long been my dilemma. One likes to tell others about enjoyable experiences, but it is easy to become a bore. Would it not be too presumptuous to think that these stories would interest any readers except those as obsessed as myself? Perhaps it is, but then there are many such readers. I know only too well that many of my fellow-men are also naturalists and treasure-hunters at heart, and often feel an urge to break out for a while and live a little closer to Nature. Perhaps, since I have been more fortunate than most in being able to fulfil that urge, my stories might encourage such people to do as I did; I can assure them that it has always been fun.

I even have the temerity to believe that my book might perhaps help in recruiting more young naturalists to biology. Increasingly exact and specialized research is being done nowadays, yet some fields of biology seem sometimes to lose touch with biology's original object—living things in their natural surroundings. Perhaps books such as this might show budding naturalists with inquiring minds that they can still play a part in biology. If so, a little note of warning might not be amiss: new recruits will have to be more systematic and more thorough than I have been. This, however, need not discourage them; even consistency and thoroughness, if not reduced to obsessions, can be fun.

ACKNOWLEDGMENTS

I would like on this occasion to express my gratitude to three friends who initiated me in the study of animal behaviour: Dr A. Schierbeek, the late G. J. Tijmstra, and Dr Jan Verwey. I further owe much to the friendship and co-operation of my co-workers, whose names are mentioned in the main text.

For permission to use Figs. 1, 3, and 13 I am indebted to the Clarendon Press, Oxford; Professor G. P. Baerends kindly allowed me to use Figs. 10, 11 and 12; for advice on Fig. 20 I am indebted to Dr E. Cullen; and Figs. 23, 24 and 25 are reproduced with permission of Dr A. Manning and Messrs E. J. Brill, Leiden.

Professor Baerends kindly read Chapter 6, and Dr J. M. Cullen read the entire manuscript and gave me the benefit of his criticism.

N.T.

CONTENTS

ILLUSTRATIONS

CURIOUS NATURALISTS

CHAPTER ONE

THE BEE-HUNTERS OF HULSHORST

As a child I thoroughly disliked insects and always suspected that, apart from crawling over one's skin, they all might bite or sting. Later, however, I spent many summers studying the habits of one of the fiercest of digger wasps, *Philanthus*, the bee-killer, and gradually I grew very fond of this insect and of many others. I owe this interest to the lovely country of Hulshorst.

In my early twenties, when I had been a zoology student for several years, I still showed a certain preoccupation with hockey, pole-jumping, skating, camping and bird-photography. My zoology professor did not altogether approve of this; in fact, he had given me up as an incipient zoologist, and I cannot altogether blame him. Yet I was a budding naturalist—of a kind—and without being aware of it I was waiting for the critical stimulus that would start me on what was to become my hobby and my job.

The summer vacation usually found my parents, with all their children and many friends, in occupation of a small cottage in one of the less densely populated areas of crowded little Holland, amid wide expanses of glacial sands which, fortunately enough, were good for nothing except, on the best parts, growing low-quality timber. Some irregular strips of fine deciduous trees such as Beech and Oak bordered a brook that wound its way through the heaths and sand dunes on its way to the Zuiderzee. Several square miles of the hilly sands had been planted with Scots Pine, but large stretches were covered with a carpet of Heather, and other parts again were just bare sand of the poorest sort. Isolated wind-blown Pines, and little dunes with a sparse vegetation of Marram Grass, broke the monotony of these arid plains.

But don't think that this country was dull—far from it! On countless long walks, at first in the summer, later in all seasons of the year, we discovered its many hidden treasures. The cool green Beech wood bordering the brook harboured such fine birds as Black Woodpeckers, Honey Buzzards and Woodcock. Quietly following the little stream, we often surprised Roedeer nibbling at fresh green shoots, or drinking in a quiet corner. In the autumn, when the fallen Beech leaves glowed like polished copper, we were delighted to find the most wonderful mushrooms; collecting the edible species gave a culinary value to our leisurely walks, and made us sympathize with the Red Squirrels who were busy nibbling at acorns or burying them.

The Scots Pine plantations, at first glance so monotonous, were full of fascinating creatures, many of them beautifully camouflaged. The larvae of the Pine Hawk Moths and several other caterpillars were living among the needles, eating away steadily, blending with their background through their wonderful pattern of green and white stripes. In the poorer parts of the woods, where the trunks were covered with lichens, a peculiar fauna of spiders, bugs and moths was found, all matching their background by their remarkably efficient colour patterns of a pale lichen green covered with black blotches.

On the greenish grey carpet of reindeer moss, soft as velvet or crisp as toast, according to the weather, the lovely orange-ochre Chanterelles glowed all through the summer and autumn—and they supplied us with many a good meal.

Flocks of tits travelled through the tops of the trees, continually whispering to keep in touch with each other. Their whispers stopped only when a sharp 'zee' of one of them warned them of an attacking Sparrow Hawk. These, us well as their bigger cousins, the Goshawks, could be found nesting in the older parts of the woods.

One dramatic winter day found us walking through these woods during a fierce blizzard. Starting as rain, then turning to sleet, then, with dropping temperature, into dry snow, huge masses of moisture were hurled down on the wood, stuck to the needles and the twigs, and finally, through their accumulated weight, began to break even the strongest branches. Above the roar of the gale, the cracking and snapping of the overloaded branches rang through the woods like so many rifle shots. Looking at this, not with the eyes of a forester but with those of the naturalist, we thought ourselves fortunate in witnessing one of those rare, critical days, so interesting to the ecologist because they may well have tremendous consequences for wild life.

Quite different were the wide heaths. In late summer, when the Heather bore its myriads of flowers, there was a purple glow over the undulating plains, fading away in the heat haze, the horizon shimmering and quivering in the hot sun. The flowers of the Heather, producing a delicious nectar, attracted innumerable insects, whose continuous buzzing was as much part of the scene as the sound of the surf is part of the sea shore. Here and there we would find rows of primitive straw bee-hives, belonging to small crofters living up to twenty miles away, who had carted them up for the short but prolific season.

On long walks one could suddenly chance upon tiny

moors, with cotton grass and many colourful plants; a family of Curlews disturbed by our intrusion would fly off, uttering their melodious calls.

Finally, even the bare sandy stretches had a charm of their own. The sparse vegetation could not manage to cover the soil entirely, and on windy days the sand was blown over the dune crests. Thus, the small hills were continually on the move, and since south-westerly winds prevail in this part of the world, the bulk of the sand was always shifting to the north-east. On the south-westerly edge of the sands there were flat, blown-out plains, whereas on the north-east side the bare sand tongues moved slowly but irresistibly forward, burying everything in their path, heath and forest alike.

On the flats the gravel, too heavy to be blown away, was left behind, and this put a stop to wind erosion. Lichens, mosses, grasses and finally Pine seedlings settled here, and gradually such plains were covered by a very open vegetation of Pine, *Calluna*, and scattered patches of lichens and mosses.

The sands, and these natural Pine woods, were not as poor as they might seem. They were lovely insect country. Thousands of Ant Lions preyed on the Red Wood Ants which abounded where isolated Oak and Birch trees supported a rich supply of insects. Pine Hawk Moths lived on the isolated Pines. Nightjars were abundant. Dung Beetles lived on the droppings of the many Rabbits; on sunny days they flew about high up in the sky, and Hobbies, those elegant little falcons, spent hours hawking them on the wing. Tiger Beetles, large predatory flies (*Asilus*) and many kinds of burrowing wasps earned a living by preying on other insects, either those living there on the sparse vegetation or others, such as the hapless Crane Flies, that strayed into the area.

We lived in that country long enough to have seen it in a great variety of circumstances and moods, and gradually every part of it acquired significance; the whole area became charged with our experiences, many of which I shall never forget. Here we had found the flint implements left by people living in this place several thousand years ago. There we had seen the lightning strike a tree, shattering its bark, and throwing large pieces of it all around. At another place, right in the middle of the sands, we had found another witness of lightning, where it had struck the ground—a 'fulgurite' or tube of molten sand. We dug the whole thing up, all 12 ft. of it, and it is now in the Leiden Geological Museum.

Some parts of the woods are forever imprinted in our memory as the haunts of the Goshawk; we spent many hours in a tree-top hide watching this formidable predator feeding its young in the large nest. And one particular corner of the sands, a blown-out, flat valley of a quarter of a mile square, is my favourite spot since it was there that I discovered *Philanthus*, the digger wasp that made me take up my first experimental study—and so helped me to gain some self-respect.

On a sunny day in the summer of 1929 I was walking rather aimlessly over the sands, brooding and a little worried. I had just done my finals, had got a half-time job, and was hoping to start on research for a doctor's thesis. I wanted very much to work on some problem of animal behaviour and had for that reason rejected some suggestions of my well-meaning supervisor. But rejecting sound advice and taking one's own decisions are two very different things, and so far I had been unable to make up my mind.

While walking about, my eye was caught by a bright orange-yellow wasp the size of the ordinary jam-loving *Vespa*. It was busying itself in a strange way on the bare sand. With brisk, jerky movements it was walking slowly backwards, kicking the sand behind it as it proceeded.

The sand flew away with every jerk. I was sure that this was a digger wasp. The only kind of that size I knew was *Bembex*, the large fly-killer. But this was no *Bembex*. I stopped to watch it, and soon saw that it was shovelling sand out of a burrow. After ten minutes of this, it turned round, and now, facing away from the entrance, began to rake loose sand over it. In a minute the entrance was completely covered. Then the wasp flew up, circled a few times round the spot, describing wider and wider loops in the air, and finally flew off. Knowing something of the way of digger wasps, I expected it to return with a prey within a reasonable time, and decided to wait.

Sitting down on the sand, I looked round and saw that I had blundered into what seemed to be a veritable wasp town. Within ten yards I saw more than twenty wasps occupied at their burrows. Each burrow had a patch of yellow sand round it the size of a hand, and to judge from the number of these sand patches there must have been hundreds of burrows.

I had not to wait long before I saw a wasp coming home. It descended slowly from the sky, alighting after the manner of a helicopter on a sand patch. Then I saw that it was carrying a load, a dark object about its own size. Without losing hold if it, the wasp made a few raking movements with its front legs, the entrance became visible and, dragging its load after it, the wasp slipped into the hole.

At the next opportunity I robbed a wasp of its prey, by scaring it on its arrival, so that it dropped its burden. Then I saw that the prey was a Honey Bee.

I watched these wasps at work all through that afternoon, and soon became absorbed in finding out exactly what was happening in this busy insect town. It seemed that the wasps were spending part of their time working at their burrows. Judging from the amount of sand excavated these must have been quite deep. Now and then a wasp would fly out and, after half an hour or longer, return with a load, which was then dragged in. Every time I examined the prey, it was a Honey Bee. No doubt they captured all these bees on the heath for all the to and fro traffic was in the direction of the south-east, where I knew the nearest heath to be. A rough calculation showed that something was going on here that would not please the owners of the bee-hives on the heath; on a sunny day like this several thousand bees fell victims to this large colony of killers!

As I was watching the wasps, I began to realize that here was a wonderful opportunity for doing exactly the kind of field work I would like to do. Here were many hundreds of digger wasps—exactly which species I did not know yet, but that would not be difficult to find out. I had little doubt that each wasp was returning regularly to its own burrow, which showed that they must have excellent powers of homing. How did they manage to find their way back to their own burrow?

At that time the excellent studies of several German zoologists, notably E. Wolf, had already proved that Honey Bees were quite good at homing, and much had already been discovered of the way they did it. But I knew that the homing abilities of solitary wasps were still unexplored, and that the observations of that master of entomology, Henry Fabre and of his followers, such as Ferton, Rau and others, rather suggested that such wasps had quite mysterious powers of finding their way back to their own burrows. But the work of these naturalists, though admirable in many ways, had always left me unsatisfied, and obviously something had to be done about this problem. Here seemed to be a first-rate opportunity. I then and there decided to tackle it.

There was another puzzling thing about these wasps. If they specialized so entirely on Honey Bees—as they seemed to do—how did they recognize them among the teeming thousands of insects of so many kinds that were feasting on the Heather? Would it be possible to find out?

My worries were over; I knew what I wanted to do. This day, as it turned out, was a milestone in my life. For several years to come I was to spend my summers with these wasps, first alone, then with an ever-growing group of co-workers, most of them zoology students of the University of Leiden. Soon we began to study other insects as well, and thus started what has come to be a tradition in our Zoology Laboratory, of organizing biological summer camps year after year at Hulshorst. We owe a great debt of gratitude to the generous owner of this lovely stretch of country, Mr A. E. Jurriaanse, who so cordially welcomed us every summer, who allowed us to go about our business (which must at times have seemed a little odd), and who gave us support of so many kinds.

When in 1949 I left Holland to settle in Oxford my colleague, Dr Jan van Iersel, took over and, while I write this, the Hulshorst work is still going strong on a considerably expanded scale.

Settling down to work, I started spending the wasps' working days (which lasted from about 8 a.m. till 6 p.m. and so did not put too much of a strain on me) on the 'Philanthus plains', as we called this part of the sands as soon as we had found out that *Philanthus triangulum Fabr.* was the official name of this bee-killing digger wasp. Its vernacular name was 'Bee-Wolf'.

An old chair, field glasses, note-books, and food and water for the day were my equipment. The local climate of the open sands was quite amazing, considering that ours is a temperate climate. Surface temperatures of 110° F were not rare, and judging from the response of my skin, which developed a dark tan, I got my share of ultraviolet radiation.

My first job was to find out whether each wasp was really limited to one burrow, as I suspected from the unhesitating way in which the home-coming wasps alighted on the sand patches in front of the burrows. I installed myself in a densely populated quarter of the colony, five yards or so from a group of about twenty-five nests. Each burrow was marked and mapped. Whenever I saw a wasp at work at a burrow, I caught it and, after a short unequal struggle, adorned its back with one or two colour dots (using quickly drying enamel paint) and released it. Such wasps soon returned to work, and after a few hours I had ten wasps, each marked with a different combination of colours, working right in front of me. It was remarkable how this simple trick of marking my wasps changed my whole attitude to them. From members of the species *Philanthus triangulum* they were transformed into personal acquaintances, whose lives from that very moment became affairs of the most personal interest and concern to me.

While waiting for events to develop, I spent my time having a close look at the wasps. A pair of lenses mounted on a frame that could be worn as spectacles enabled me, by crawling up slowly to a working wasp, to observe it, much enlarged, from a few inches away. When seen under such circumstances most insects reveal a marvellous beauty, totally unexpected as long as you observe them with the unaided eye. Through my lenses I could look at my *Philanthus* right into their huge compound eyes; I saw their enormous, claw-like jaws which they used for crumbling up the sandy crust; I saw their agile black antennae in continuous, restless movement; I watched their yellow, bristled legs rake away the loose sand with such vigour that it flew through the air in rhythmic puffs, landing several inches behind them.

Soon several of my marked wasps stopped working at their burrows, raked loose sand back over the entrance, and flew off. The take-off was often spectacular. Before leaving they circled a little while over the burrow, at first low above the ground, soon higher, describing ever widening loops; then flew away, but returned to cruise once more low over the nest. Finally, they would set out in a bee-line, fifteen to thirty feet above the ground, a rapidly vanishing speck against the blue sky. All the wasps disappeared towards the south-east. Half a mile away in that direction the bare sands bordered upon an extensive heath area, buzzing with bees. This, as I was to see later, was the wasps' hunting area.

The curious loops my wasps described in the air before leaving their home area had been described by other observers of many other digger wasps. Philip Rau had given them the name of 'locality studies'. Yet so far nobody had proved that they deserved that name; that the wasps actually took in the features of the burrow's surroundings while circling above them. To check this if possible was one of my aims—I thought that it was most probable that the wasps would use landmarks, and that this locality study was what the name implied. First, however, I had to make sure that my marked wasps would return to their own holes.

When my wasps had left, there began one of those periods of patient waiting which are usual in this kind of work. It was, of course, necessary to be continually on the look-out for returning wasps; at the same time it was tempting to look round and watch the multitude of other creatures that were busy on the hot plains. For I soon discovered that I had many neighbours. First of all, there were other diggers about. Among the *Philanthus* burrows there were some that looked different—the sand patches were a little larger and less regular. These belonged to the largest of our digger wasps, the fly-killing *Bembex*, that is almost the size of a Hornet. Buzzing loudly, flying with terrific speed low over the ground, these formidable wasps dashed to and fro, and it took me a long time before I saw more of them than a momentary sulphur-yellow flash. A Leafcutter Bee was coming home carrying its 'wall paper', a neat circular disc cut out of a rose leaf. Its burrow, scarcely visible, was in the carpet of dry moss just beyond the *Philanthus* settlement. Robber Flies (*Asilus crabroniformis*) whizzed past, catching flies and other insects in the air. Sometimes they would make a mistake and attack *Philanthus*. A brief but furious buzzing struggle, and the two fell apart, *Asilus* darting off to find a less petulant prey, *Philanthus* returning to its burrow. Small grasshoppers came walking by, greedy and single-minded, devouring one grass shoot after the other, or spending hours chirping their song, or courting, with pathetic perseverance, seemingly unconcerned females.

On some days we saw endless processions of migrating Cabbage White butterflies crossing the plains in a continuous stream of scattered formations, usually going north-west. Hobbies came and levied their toll on them, as well as on the many dragonflies and the clumsy Dung Beetles. Bumblebees often zoomed past on their long, mysterious trips and, like the Cabbage Whites, would show their interest in nectar-filled blue flowers by alighting on anything blue in our equipment.

In August the monotony of the deep blue sky might be broken by a lonely Osprey, flying in from his fishing grounds north of us—the coastal waters of the Zuiderzee —to settle in the crown of an old Pine, there to dream away the hours of digestion. Or a group of lovely, black-and-white Storks, on migration to Africa, might come sailing by, stopping in the rising air over the hot sands to soar up in wide circles, higher and higher, until they resumed their glide south in search of the next 'thermal'. Thus sitting and waiting for the wasps was never dull, if only one kept one's eyes open.

To return to my marked wasps: before the first day was over, each of them had returned with a bee; some had returned twice or even three times. At the end of that day it was clear that each of them had its own nest, to which it returned regularly.

On subsequent days I extended these observations and found out some more facts about the wasps' daily life. As in other species, the digging of the large burrows and the capturing of prey that served as food for the larvae was exclusively the task of the females. And a formidable task it was. The wasps spent hours digging the long shafts, and throwing the sand out. Often they stayed down for a long time and, waiting for them to reappear, my patience was often put to a hard test. Eventually, however, there would be some almost imperceptible movement in the sand, and a small mound of damp soil was gradually lifted up, little by little, as if a miniature Mole were at work. Soon the wasp emerged, tail first, and all covered with sand. One quick shake, accompanied by a sharp staccato buzz, and the wasp was clean. Then it began to mop up, working as if possessed, shovelling the sand several inches away from the entrance.

I often tried to dig up the burrows to see their inner structure. Usually the sand crumbled and I lost track of the passage before I was ten inches down, but sometimes, by gently probing with a grass shoot first, and then digging down along it, I succeeded in getting down to the cells. These were found opening into the far end of the shaft, which itself was a narrow tube, often more than 2 ft. long. Each cell contained an egg or a larva with a couple of Honey Bees, its food store. A burrow contained from one to five cells. Each larva had its own living room-cum-larder in the house, provided by the hard-working female. From the varying number of cells I found in the nests, and the varying ages of the larvae in one burrow, I concluded that the female usually filled each cell with bees before she started to dig a new cell, and I assumed that it was the tunnelling out of a new cell that made her stay down for such long spells.

I did not spend much time digging up the burrows, for I wanted to observe the wasps while they were undisturbed. Now that I was certain that each wasp returned regularly to her own burrow, I was faced with the problem of her orientation. The entire valley was littered with the yellow sand patches; how could a wasp, after a hunting trip of about a mile in all, find exactly her own burrow?

Having seen the wasps make their 'locality studies', I naturally believed that each female actually did what this term implied: take her bearings. A simple test suggested that this was correct. While a wasp was away I brushed over the ground surrounding the nest entrance, moving all possible landmarks such as pebbles, twigs, tufts of grass, Pine cones, etc, so that over an area of 3–4 square metres none of them remained in exactly the same place as before. The burrow itself, however, I left intact. Then I awaited the wasp's return. When she came, slowly descending from the skies, carrying her bee, her behaviour was striking. All went well until she was about 4 ft. above the ground. There she suddenly stopped, dashed back and forth as if in panic, hung motionless in the air for a while, then flew back and up in a wide loop, came slowly down again in the same way, and again shied at the same distance from the nest. Obviously she was severely disturbed. Since I had left the nest itself, its entrance, and the sand patch in front of it untouched, this showed that the wasp was affected by the change in the surroundings.

Gradually she calmed down, and began to search low

over the disturbed area. But she seemed to be unable
to find the nest. She alighted now here, now there, and
began to dig tentatively at a variety of places at the ap-
proximate site of the nest entrance. After a while she
dropped her bee and started a thorough trial-and-error
search. After twenty-five minutes or so she stumbled
on the nest entrance as if by accident, and only then
did she take up her bee and drag it in. A few minutes
later she came out again, closed the entrance, and set
off. And now she had a nice surprise in store for
me: upon leaving she made an excessively long 'locality study': for fully two minutes she circled and
circled, coming back again and again to fly over the
disturbed area before she finally zoomed off.

I waited for another hour and a half, and had the
satisfaction of seeing her return once more. And what
I had hoped for actually happened: there was scarcely
a trace of hesitation this time. Not only had the wasp
lost her shyness of the disturbed soil, but she now
knew her way home perfectly well.

I repeated this test with a number of wasps, and their
reactions to my interference were roughly the same each
time. It seemed probable, therefore, that the wasps
found their way home by using something like landmarks in the environment, and not by responding to
some stimulus (visual or otherwise) sent out by the nest
itself. I had now to test more critically whether this was
actually the case.

The test I did next was again quite simple. If a wasp
used landmarks it should be possible to do more than
merely disturb her by throwing her beacons all over the
place; I ought to be able to mislead her, to make her
go to the wrong place, by moving the whole constella-
tion of her landmarks over a certain distance. I did this
at a few nests that were situated on bare sandy soil and
that had only a few, but conspicuous, objects nearby,

such as twigs, or tufts of grass. After the owner of such
a nest was gone, I moved these two or three objects
a foot to the south-west, roughly at right angles to the
expected line of approach. The result was as I had hoped
for and expected, and yet I could not help being surprised as well as delighted: each wasp missed her own
nest, and alighted at exactly the spot where the nest
'ought' to be according to the landmarks' new positions!
I could vary my tests by very cautiously shooing the
wasp away, then moving the beacons a foot in another
direction, and allowing the wasp to alight again. In
whatever position I put the beacons, the wasp would follow them. At the end of such a series of tests I replaced
the landmarks in their original position, and this finally
enabled the wasp to return to her home. Thus the tests
always had a happy ending—for both of us. This was
no pure altruism on my part—I could now use the wasp
for another test if I wished.

When engaged in such work, it is always worth observing oneself as well as the animals, and to do it as
critically and as detachedly as possible—which, of course,
is a tall order. I have often wondered why the outcome
of such a test delighted me so much. A rationalist would
probably like to assume that it was the increased pre-
dictability resulting from the test. This was a factor of
considerable importance, I am sure. But a more important factor still (not only to me, but to many other
people I have watched in this situation) is of a less dig-
nified type: people enjoy, they relish the satisfaction of
their desire for power. The truth of this was obvious, for
instance, in people who enjoyed seeing the wasps being
misled without caring much for the intellectual question
whether they used landmarks or not. I am further convinced that even the joy of gaining insight was not often
very pure either; it was mixed with pride at having had
success with the tests.

To return to the wasps: next I tried to make the wasps use landmarks which I provided. This was not only for the purpose of satisfying my lust for power, but also for nobler purposes, as I hope to show later. Since changing the environment while the wasp was away disturbed her upon her return and even might prevent her from finding her nest altogether, I waited until a wasp had gone down into her nest, and then put my own landmarks round the entrance—sixteen Pine cones arranged in a circle of about eight inches diameter.

The first wasp to emerge was a little upset, and made a rather long locality study. On her return home, she hesitated for some time, but eventually alighted at the nest. When next she went out she made a really thorough locality study, and from then on everything went smoothly. Other wasps behaved in much the same way, and next day regular work was going on at five burrows so treated. I now subjected all five wasps, one by one, to a displacement test similar to those already described. The results, however, were not clear-cut. Some wasps, upon returning, followed the cones; but others were not fooled, and went straight home, completely ignoring my beacons. Others again seemed to be unable to make up their minds, and oscillated between the real nest and the ring of cones. This half-hearted behaviour did not disturb me, however, for if my idea was correct —that the wasps use landmarks—one would rather expect that my tests put the wasps in a kind of conflict situation: the natural landmarks which they must have been using before I gave them the Pine cones were still in their original position; only the cones had been moved. And while the cones were very conspicuous landmarks, they had been there for no more than one day. I therefore put all the cone-rings back and waited for two more days before testing the wasps again. And sure enough, this time the tests gave a hundred per cent preference for the Pine cones; I had made the wasps train themselves to my landmarks.

Fig. 1. A homing test. Philanthus returns to the displaced pine cones and fails to find her burrow.

The rest of this first summer I spent mainly in consolidating this result in various ways. There was not much time to do this, for the season lasts only two months; by the end of August the wasps became sluggish, and soon after they died, leaving the destiny of their race in the hands of the pupae deep down in the sand, which were to lie there dormant until next July. And even in this short summer season the wasps could not work steadily, but were active on dry sunny days only—and of these a Dutch summer rarely supplies more than about twenty in all.

However, I had time to make sure that the wasps relied for their homing mainly on vision. First, I could cut off their antennae—the bearers of delicate organs of smell, of touch and of other sense organs—without at all disturbing their orientation. Second, when, in other tests, I covered the eyes of intact wasps with black paint, the wasps could not fly at all. Removing the cover of paint restored their eyesight, and with it their normal behaviour. Furthermore, when I trained a wasp to accept a circle of Pine cones together with two small squares of cardboard drenched in Pine oil, which gave off a strong scent, displacement of the cones would mislead the wasps in the usual way, but moving the scented squares had not the slightest effect. Finally, when wasps used to rings of cones were given, instead of cones, a ring of grey pebbles a foot from the nest, they followed these pebbles. This can only have been due to the pebbles being visually similar to the cones.

This was about as far as I got in that first summer. I had had my first thrills as an experimenter. However simple my observations and tests had been, they had led me to some real discoveries. I had tasted that particular brand of triumph that is the reward of true exploration. From now on my plans for the summer vacations for years to come were made: I had to keep going back to Hulshorst and find out more about the bee-killers. As it turned out, this decision fixed the summer plans of my wife, of all our children, and of several 'generations' of students as well: for the field work on the sandy plains went on for many years, even after the bee-killers, unusually abundant in these first years, had fallen back to their normal status of a relatively rare species, and so forced us to abandon their study.

CHAPTER TWO

ARCTIC INTERLUDE

Back from my first season's field work on the homing of
the digger wasps, I got an opportunity which few naturalists ever get. A small Dutch expedition was going to carry out a full year's meteorological observations
on the ice-bound coast of East Greenland, and by an
extremely lucky coincidence I was allowed to attach myself to this group to do zoological research. When I
heard of it, I did not hesitate a moment. To throw
away such a chance would have seemed utter madness
to me. To see the Arctic and its wild life, the pack ice,
icebergs, to live among the Eskimos—it surpassed my
wildest dreams.

In the course of the winter 1931–32 I made my plans.
The first complication was that I was already engaged
to be married. Neither of us had any doubts as to the
right course to take—we would have to go together. How
that plan was put into action need not concern us here,
but to our immense joy it was so.

Then, I had to produce a doctor's thesis dealing with
my work on *Philanthus*. I am afraid that this little matter was dealt with in a rather cursory way. Surprisingly
enough, however, everything went smoothly. We were
married the day after and, putting off our honeymoon
for the time being, we got down to working out our plans
in detail, purchasing and trying out our Arctic equip-

ment, practising with a folding canoe, learning Danish, and a hundred and one other things. In July we left our crowded little country and after a brief stay in Copenhagen, where we joined the meteorological party of four, we sailed on a rainy day, in the 800 ton four-master *Gertrud Rask*.

For us greenhorns it was another stroke of luck that we sailed on the same ship as four young but already highly experienced British explorers—Gino Watkins, Freddy Spencer Chapman, John Rymill, and Quentin Riley. Our theoretical training of life in the Arctic, based mainly on Nansen's *First Crossing of Greenland* and *Farthest North*, and Stefánsson's *My Life with the Eskimo* and *The Friendly Arctic*, could now be supplemented by the good advice these very competent men were kind enough to give us. Little did we apprehend that the gifted leader, Watkins, would soon meet his tragic death in the country where he had worked so happily and so successfully.

In about ten days we crossed the grey North Atlantic, slowly ploughing our way over mountainous seas under a leaden sky, the wooden ship moaning and creaking as it climbed and fell over the ever-charging swell, and continually escorted by the sailing Fulmars. Soon after touching Seydisfjord, in North-east Iceland, we struck the first fields of scattered pack ice, big, pale green floes lazily rocking on the swell. Here and there we met a remnant of an iceberg, oddly out of place right there in the open ocean, and exposed to the assaults of the waves of warm surface water. Yet even these ruins, with their weird shapes, battered by the spray, were of an indescribable majesty.

Zigzagging, banging into the ice, backing, turning, and trying in another direction, we slowly made our way into the pack. Once we were inside we were out of reach of the Atlantic swell, and for the first time my poor wife, as bad a sailor as I have ever seen, could come on deck and receive her reward for ten days' suffering: ice floes glittering in the sun as far as we could see. It is impossible to describe the impact this first encounter with the Arctic made on us. Like so many before us, we fell under its spell at once. For hours and hours we stood and watched, drinking in the mood of the pack, restful, aloof and somehow threatening. Yet the abundance and carefree behaviour of the wild animals, which were plentiful, reassured us and attracted us. Little Auks and Guillemots were swimming in the open lanes and pools, groups of Bladdernose Seal were basking on the ice, sometimes with elegant little Ivory Gulls in attendance.

Soon we met dense fog, and navigation became difficult, progress slow. We did not get out of it until, two days later, we made a superbly romantic landfall at Scoresby Sound just when the midnight sun broke through the low-hanging rain clouds, suddenly revealing the land of our dreams: rugged peaks with vast, undulating promontories at their base, bordering upon the bluish-grey fjord, which was covered with scattered pack ice and towering icebergs. Zigzagging through the pack a number of Eskimo hunters in their elegant kayaks, cutting through the smooth surface like so many dolphins, swarmed out to meet us. Slowly, gigantic in comparison with the tiny kayaks, the ship entered the shallow bay, Amdrup's Havn, and dropped anchor. We were welcomed by the entire population of the tiny settlement. The church bell chimed, shots were fired, the huskies howled.

Next morning, refreshed after a few hours' sleep, we faced an entirely different Greenland. The sky had cleared and the vast fjord was bathed in sunlight. The water was a deep blue, the pack ice and the magnificent bergs were glowing in fascinating shades of green, blue and violet. The red granite mountains, their valleys and gorges filled with soggy snowfields, their base patched with mats of bronze-green vegetation, glowed in the low arctic sun. The air was crisp, yet in the sun it was delightfully warm.

We spent five days in Scoresby Sound, which we used for establishing our first contacts with the East Greenland Eskimo, for trying out some of our equipment, and for making long trips through the country. The views of the 25 mile wide fjord were overpowering; the vastness of the country as it lay there, solidly, row after row of mighty mountains, fading gradually in the very slight haze until, 50 miles or more away, they seemed to dissolve and to fuse with the fjord and the sky; the distant icebergs losing themselves in fantastic mirages on the horizon, the peculiar transparency of the atmosphere—it was all of an indescribable beauty.

After this brief stay we weighed anchor and proceeded south along the coast, keeping several miles out, and steering well away from the many icebergs. We had brilliant sunshine at first, so that we had a good view of the high alpine mountains crowned with the mighty ice cap, from which huge glaciers flowed down into the sea. Now and then we had to cross fields of pack ice. Soon, however, we struck dense fog, which held us in its grip until we reached Angmagssalik Harbour, a sheltered bay in the small Tassiussaq fjord, at about 67° N. Here again we were welcomed by a crowd of kayakmen, swarming out in far greater numbers than in Scoresby Sound, and then escorting the ship as it glided into the snug little corner of the fjord where the main settlement of Greenland's East coast was: a dozen or so wooden houses, built by the Danes, round which the low Eskimo huts were grouped. The settlement was laid out at the mouth of a small brook that came running down the hills, and was surrounded by low mountains, with some 2000 ft. peaks behind. The opposite side of the fjord was formed by an impressive array of peaks about double this height.

Tassiussaq, as this settlement was called by the Eskimos, was the main and, until ten years previously, the only settlement on the East coast. A tribe of about 800 Eskimos, isolated from the West coast since the 15th century, discovered in 1883, and in contact with the white man since 1910, lived in scattered groups on the shores of the three large fjords, Sermilik, Angmagssalik fjord, and Sermiligâq. For the annual occasion of the arrival of the ship most of these Eskimos had gathered in Tassiussaq. A Danish governor, a Danish wireless operator and a West Greenland missionary were living in Tassiussaq, with a few native families. All the others present were staying just for a few days and would soon leave for their tiny villages 10 to 75 miles away.

We were indeed lucky to have our headquarters here on the East coast among a relatively little-touched tribe of one of the most fascinating peoples on earth!

Soon after our arrival, when our supplies and equipment had been unloaded, we settled down to work. The plans of my wife and myself were, as usual in enterprises of this kind, various. First, we were to make as complete a collection of ethnographica as possible, for a Museum in Holland. The Angmagssalingmiut, as these people called themselves, show, in their isolation, several typical traits, and although Copenhagen could boast the excellent Thalbitzer collection very little else had ever been collected, and our opportunity for obtaining more material might well be one of the very last possible.

Further we were to make botanical and zoological collections, though this was definitely to be a sideline. Our main task would be a study of the behaviour of some typically arctic animals. To carry out this programme successfully we would have to travel and to live far from the base for most of the time. We would have to master the technique of arctic travel of various kinds, to look around before deciding where to do our field work and, since we knew that we would have to learn a great deal from the Eskimos, we had to acquire some knowledge of their language. This programme kept us pretty busy right from the beginning, and of the many experiences we had in this year I can select only a few for this narrative.

The arctic summer is short, and after we had spent a week establishing ourselves at our base there was only just time to make some reconnaissance trips through the district. We left on our first ten-day trip on August 1st, our large folding canoe loaded to the gunwale with our camping gear and other equipment, food, paraffin for the Primus stove, shotgun, etc. Although Sermilik was full of ice, mainly icebergs from the glaciers at its head and their débris, it was navigable for tiny craft such as ours, which could slip through narrow lanes and could be hauled over the ice if we got trapped. The other fjords were free of ice except for scattered icebergs. We paddled or sailed by day and camped in a small tent, usually on an island—we did not worry about water since the ice in the fjord was fresh, coming from the inland ice. New snow began to show on the highest peaks, down to about 2,000 ft., but at sea level the temperature was still quite pleasant.

Travelling inland from Tassiussaq we soon got away from the barren coast and reached the 'inner fjord zone' where, owing to shelter and the absence of the coast-bound summer fogs, we found luxuriant vegetation. The climax of this vegetation seemed to be formed by the 'forests' of knee-high Dwarf Birch and Sallow. It was obvious that we could expect the richest animal life in these parts. It was true that large plant-eating mammals such as Musk Ox and Arctic Hare were absent in the

Angmagssalik area, although Reindeer were once known to have been here. Lemmings were not present either. Consequently there were no Stoats, nor Snowy Owls, nor Long-tailed Skuas. But the sea life was rich and seals were plentiful. Brünnich's and Black Guillemots abounded and of the land birds we saw Raven, Snow Bunting, Lapland Bunting, Wheatear and Ptarmigan in large numbers.

We were particularly interested in the Snow Bunting, which we were planning to study closely next spring, and the primary aim of our trip was to find a suitable site where this lovely pied bird was common and which was not too far from an Eskimo settlement. We found the ideal place in Torssukátaq, a small side fjord of the vast Angmagssalik fjord, about 4 miles long and 40 miles from the Atlantic coast. Here the Kûngmiut lived ('the people of the river'), in a village consisting of 12 winter houses.

Here, as in the other settlements we visited on our trip, the Eskimos were busy rebuilding their winter houses. The picturesque sealskin tents in which they had lived all through the summer would provide insufficient shelter for the winter. The walls of these houses, 2–3 ft. thick and built of turf and slabs of stone, had been left standing. Now the roof beams were put up and covered with skins, turf and stones. The cracks in the walls were mended, the blubber lamps and household goods carried in. The skin-covered rowing boats were laid up for the winter; fastened upside down on strong, high stands they were turned into excellent larders, natural, well aerated refrigerators, out of reach of the dogs.

On this first journey we visited as many Eskimo families as we could and began our study of their language. We knew a few words which we had been taught by Spencer Chapman and some of the Eskimos knew one or two Danish words, so with these as a start we tackled

this job. And what fun it was! Our hosts were just as excited as we were and, while we were sharing their meals
of fresh or smoked Char, cooked seal meat and blubber-
washed crowberries, we stammered and gestured away,
raising our eyebrows in puzzled inquiry so often that we
strained our scalp muscles, and together bursting into
hearty laughter every few minutes. The Eskimos' delightfully naïve *joie de vivre* was extremely contagious.

We could soon make ourselves understood in simple
matters and while we were in Kûngmiut we succeeded
in making arrangements for settling there in the spring.

After we had made a few more reconnaissance trips
winter set in, at first with sleet and light frost, but soon
with long snow storms and spells of biting cold. In calm
spells the fjords froze over, but gales broke up the ice
again and again. It lasted till January before new fields
of pack ice arrived from the north; they damped the
Atlantic swell and allowed the fjords to freeze over
firmly. All through the autumn the snow piled up higher
and higher and was swept into tightly packed banks
by the gales. Already in November the surface was
level with the roof of our house. The days grew shorter
and in December we had a few brief hours of twilight
round midday. We were too far south for a true polar
night and, in fact, it is only the presence of the inland
ice and the existence of the southward ocean current
coming from the Polar basin that makes South-east
Greenland such a barren and truly Arctic region.

On moonlit nights the light conditions were very much
better than by 'day', and we made many nocturnal trips
in the country, silently gliding on our skis over the soft
snow, marvelling at the soft silvery light playing over
the magnificent snow-clad mountains, and at the spectacular displays of Northern Lights. The light was so
good on some of these nights that it allowed us to hunt
and many a Ptarmigan found its way to our meat pots.

We now spent many hours a day tackling the Eskimo language in earnest, and practising our knowledge on the families living in nearby settlements. The Eskimo language, related, I am told, to those of the North American Indians, has an extremely complicated grammar. It was never written until a German Herrnhutter missionary, Kleinschmidt, succeeded in fitting our alphabet to it, with some minor modifications. His system is now widely applied and seems to be completely satisfactory.

We mastered this part of our task with little difficulty. What caused us great trouble was the construction of the sentences. These often start with the root of a main word, at the end of which the roots of secondary words were added, one after the other, and only the final ending was conjugated or declined. For instance: *igdloo* is 'a house'; *igdloqarpunga* is 'I have a house, I dwell' (from *igdloo* and *qarpoq*=he has, in the first person singular). *Nanoq* is 'a bear'; *nanorpoq* is 'he kills a bear'. *Qingaq* means 'a narrow fjord'; *qingorssuaq* means 'a long narrow fjord'. A short sentence: *oqalugpagdlarâtit*='you are talking much too much' (from *oqalugpoq*=he talks, *pagdlârpoq*=he is or he does too much, and *raoq*= much, very, in the second person singular). With great ease verbs are made out of nouns; *goddagpoq* is 'he greets' and is made by adding the suffix *poq* to the Danish *Goddag*; and the children were often told '*Goddagniaritse!*' ('Say good day!') when we arrived in a village. As in other languages much depended on the sequence in which the words followed each other.

In spite of strenuous work we did not succeed in this one winter in approaching anything like mastery of the language, but we acquired a satisfactory working knowledge of it, and at the end of the year two Dutch friends, who joined us for some weeks, got the impression that we spoke it fluently, though actually we were far from it. Our conversation ran smoothly enough, but it was a horrid 'pidgin eskimo'. The wonderful courtesy of the Eskimos made them do their utmost to speak to us in a language we could understand, and they soon used our jargon—thus, with the best of intentions, completely thwarting our attempts at improvement.

That winter we spent much more time with the Eskimos than with our Dutch friends. We made numerous one-day hunting and fishing trips with our neighbours, ate their food, and spent long evenings in their houses, studying the way they made their harpoons and other weapons, listening to their folklore, watching (and joining in) their dances, and acquiring a great deal of useful practical knowledge, which was to be of vital importance to us later when we started travelling on our own.

In January, after a few practice trips to Kulusuk across the mouth of Angmagssalik fjord, we left, with a small caravan of seven heavily loaded sledges, for Kûngmiut, our headquarters-to-be. The days had just started lengthening, although temperatures were still falling. The country was by now deeply covered in snow, the pack ice was frozen solidly to the land and extended for many miles into the ocean; open water was visible only from the tops of the mountains, 2000 ft. high, from where, on clear days, we could see landmarks more than 70 miles distant.

Travelling by sledge over this country was strenuous but exhilarating work. We were in excellent training and thought nothing of running alongside the sledge for twenty minutes at a stretch, urging the dogs on and helping them over difficult parts. Where the ice was smooth or the snow hard we could take a rest now and then by sitting on the sledges whenever we got out of breath, but where the going was hard we used to help the dogs. We had to cross several passes of about 800 ft. high, sweating our way up step by step and, more difficult, keeping the dogs and the heavy loads in check on the treacherous way down. I can still blush at my incompetence when I think back to one scene which I could not forget if I would, when one sledge got out of control, overtook its dogs, ran over the poor wretches one by one, and dragged them at terrific speed, over snow, rock and ice, all the way down, until it ended on its side on the fjord ice below, the exhausted and battered dogs strewn all round. As usual on such occasions, our Eskimo friends roared with laughter, and after half an hour's hard work we were ready to continue.

Thanks to excellent conditions, we did the 50 miles to Kûngmiut in one day. The inner part of the fjord was frozen over smoothly, and the cold, clear weather made even the salt water ice suitable for skating. My wife and I lost no time trying our skates out and soon, to the intense delight of the Eskimos, we were indulging in our national sport, cruising leisurely well ahead of the dogs, to whom our presence in front was just as powerful a spur as the hare is to the greyhounds.

Night fell and a nearly full moon guided us, throwing brilliant light on the long fjord, which wound its way inland between the towering peaks. Eventually tiny yellow light specks showed up at the foot of the mountains ahead, and soon after we pulled up in the village of Kûngmiut, a cluster of tiny houses covered by the massive blanket of snow, snugly sheltered at the foot of steep mountains, which protected it from the impact of the dreaded *Pidirrâq*, the winterly gales from the north-west that occasionally came roaring down from the ice cap.

We were given a tremendous welcome by the whole population and were soon enjoying a meal of steaming hot seal meat in the house of our host, Kârale, with whom we were going to stay until May, when we would move out into our own tents.

Now our true life in the Arctic began. Kûngmiut was an untouched Eskimo settlement far from the temptations of the metropolis of Tassiussaq. It was situated near rich sealing fjords and the people were still living much as their ancestors had done for centuries. The only innovations were skis, woollen sweaters (which, however, had not entirely ousted the original seal- and bear-skin clothes), and rifles, which were used in winter, though not in summer, when seals were still hunted by harpoon. How near these people still were to the Stone Age was suddenly forced home to us when we found stone implements still in the men's toolboxes; many of them had belonged to their fathers and grandfathers.

We were now entering our 'advanced course'. Almost every day we were out in the field, often joining the hunters on long trips to their distant hunting grounds, learning their techniques, observing the ice, the weather and the animal life, and making preparations for our spring work.

When sledging, and also when staying in the village, we had a wonderful opportunity for studying the behaviour of the Huskies. The ancestry of these dogs is only vaguely known. They are certainly closely related to Wolves and what we saw of their behaviour confirmed this. Their voice is similar to that of Wolves and the outbursts of their long-drawn, high-pitched howls were by now a familiar sound to us, coming to mean home. There were 35 hunters in Kûngmiut, of whom about 20 possessed a full pack of 6–10 dogs. Each pack lived round their owner's house and in winter, when they were well fed, they did not stray from home.

The most interesting aspect of their behaviour was the fact that these packs defended group territories. All members of a pack joined in fighting other dogs off, the males being more aggressive than the females. This tendency to join forces when attacking strange dogs was the more striking since within each pack relations were far from friendly. Yet there was no system of individual territories; each dog had complete freedom of movement on the territory of its own pack. The frictions within the pack arose over matters of social rank. As in so many social animals and birds each dog knew its companions individually. Each of them knew exactly, having learnt by (sometimes bitter) experience, which ones he had to avoid and which he could dominate without fear of retaliation.

In most packs the top dog was a strong male; next to him was his favourite wife. Such a leader claimed possession of everything he wanted; a growl or even a frown was enough to make the others withdraw or even bolt. At the bottom was the weakest dog, male or female; this poor wretch led a miserable life, walking or often crawling round at a safe distance from the others, its tail between its legs, glancing anxiously at this or that dog, never daring to take any food in which any of the others was interested, and even being in a state of terror when, on rare occasions, he could gather courage to approach a female in heat. The only time when such a dog seemed to be reasonably free of inhibitions was when he joined in the pack's attack on a trespassing neighbour.

The clashes between neighbouring packs were extremely interesting to watch. If they met at the boundary between their two territories, where the issues were even, neither group attacked. The males, and more particularly the leaders, growled at each other, and every now and then they lifted a leg and urinated—'planting a scent flag' as it can be called, for this is a means of staking out a territory and advertising it by smell. The state of tension in these strongly aroused, yet inhibited, champions also showed itself in acts which, in their similarity to human behaviour, were a source of endless amusement to us: they took it out of their own pack and the unfortunate dog of low rank who happened to come too near was growled at, or even severely mauled.

The funniest of these clashes occurred when a pack found neighbours trespassing and could chase them off because they were 'in their rights'. This happened often to the leader dog and one of his lower-ranking buddies belonging to the Eskimo next to us. These two dogs were irresistibly attracted to our refuse bin and whenever they got the chance they would trespass and have a go at it. They had a very bad conscience, however, and were always casting furtive glances over their shoulders. As soon as our dogs turned up, the intruders turned tail. The subsequent chase began in silence, but once the intruders reached their own ground they began to yap in the typical, resentful way of dogs in a state of powerless fury, while ours barked the self-asserted challenge of the rightful owners. We knew what would happen next: the beaten leader suddenly turned upon his weaker brother and for half a minute or so would gave him a pitiless strafing, and the growls of the leader, mixed with the yelping of his unfortunate companion, would alert all the dogs of the village. A collective howling concert was the usual conclusion of this drama.

The Eskimos had some knowledge of these social relationships among the dogs. They knew that a strong dog could not be made to get on with those just a little lower in the social scale. In Greenland, there is rarely deep and at the same time soft snow because of the frequent gales and therefore there is no need for using the Canadian single-track arrangement in sledging. Greenland dogs pull fan-wise; this is safest on treacherous sea ice, as the dogs' weight is spread over a large area. I do not know whether in Canada the dog in front, often called the leader, is also the tyrant of the pack. In Greenland the leader often wants his favourite wife at his side, and woe betide the dog who happens to come between. Most Eskimo hunters knew something of this. In other respects they showed just as little understanding as so many dog owners in other countries. When there was a clash between two dogs they tried to stop the fighting by whipping the aggressor—usually the one higher in the scale. This, however, made him take it out of his enemy with increased bitterness. Apparently the best thing was to let the dogs settle such matters for themselves, or even to add a little weight to the leader's arguments by hitting the lower dog; it may not seem fair, but it often made the latter 'know its place'—with the result that the work was done better.

We were fascinated by the group territories. They occur in other species as well, but I knew of no scientific studies of it. We found two facts of considerable interest. First, the place and the size of the group territory were affected by food. We could extend a pack's territory by providing food outside their original territory. This gave us the link with Wolves. It is known that a pack of Wolves in a natural state roams over a wide area and that, as a rule, such an area is hunted by only one pack —although some trespassing does occur. The best studies on the behaviour of Wolves have been made by Schenkel in the Bâle Zoo. His Wolves behaved so similarly to our sledge dogs that I am convinced that it is hostility that keeps the Wolf-packs spaced-out in Nature.

The dogs' territories were so small because they did not need to roam far for their food; their master provided it at his front door. The fact that 'our' dogs, once they had befriended us, followed us so readily to new sites where we provided food, whereas dogs of other packs did not, is an additional fact of interest. From the way our dogs treated us we were sure that we were, to them, super-leader dogs, and I believe that when a leader takes his pack to a new hunting area they will follow him. We could make our dogs invade other packs' territories to a certain extent by sheer authority; the strange dogs were afraid of us as super-leaders, and our dogs knew it.

A leader dog is not merely feared by others; he is also, in a sense, respected. By this I mean that the other dogs are not only afraid of him, they are attracted as well, and they also tend to join the leader in whatever he is doing, following his initiative. Translated into terms of human conduct, I think dogs meet their leader with a mixture of fear, affection, compliance and respect—and this description also comes, I think, very near to characterizing a dog's attitude to his master.

A second point of interest was the fact that immature dogs don't join in the defence of the territory. And it always struck us that half-grown dogs did not learn to avoid strange territories either. As long as they were very young they stayed with their mother, and so were prevented from trespassing by their attachment to her. But later, when they went out on their own, they frequently trespassed in different territories. They just went where their fancy took them; and were chased away whenever they intruded. Yet they just yelped and ran, without ever learning to avoid such areas. We followed the behaviour of two young males carefully and found, to our surprise, that when they were about eight months old they suddenly began to join their pack in fights with their neighbours. In the very same week their trespassing upon other territories became a thing of the past. And it was probably no coincidence that in that same week both made their first attempts to mate with a female in their own pack.

What struck us particularly was the sudden mastering of this particular learning task (avoiding strange territories) just when their fighting and mating behaviour appeared. There was not a sudden development of their learning capacities in general, because they had learned a great number of things before; it was a case of learning in one special situation, which seemed to be impossible to the dogs as long as their tendencies to defend the group territory and to mate were still dormant. This opened my eyes to the possibility that learning, and perhaps other higher mental abilities, might be very much more dependent on 'mood' or internal condition than we often realise; this still seems to me a problem which deserves much closer study than has been devoted to it.

The Eskimos treated their dogs badly and at times downright cruelly. This might perhaps be expected in a hunting tribe. But they also managed their teams incompetently. In West Greenland there seems to be a tradition of skilful training, and a West Greenlander we saw at work with his team was clearly superior to the East Greenlanders. With very little actual whipping he controlled his well-trained pack, whereas East Greenlanders had to reinforce commands by numerous blows. Some of them were good at giving just a slight reminder to a dog who needed it, throwing the six yard leash forward to exactly the place they wanted to touch, and touching it lightly; but many hunters were too rough, and I once saw the tip of an ear flicked off completely by a needlessly fierce lash. Some dogs had lost an eye in this way. No doubt as a consequence of this, most dogs were ill-tempered and always ready to bite, but dogs that were treated well developed good characters on the whole.

It was very amusing to see dogs react to their masters coming out with their harnesses for the first time in autumn. At the sight of him the adult dogs went wild with joy. They danced round their master, wagging their tails and trying to wedge their heads into the harnesses. The young dogs joined in the excitement, but did not know what to do; they reminded us of young boys laughing over-loud with their older friends about jokes they do not understand.

The deep snow made it possible, and indeed necessary, for us to use skis from October until well into June.

Because of the danger of breaking through the sea ice which, though often 3 ft. thick, was treacherous where tidal currents gnawed it away underneath, we used only loose straps for our feet so that we could kick ourselves free if necessary. This caused a considerable loss of manoeuvrability and took much of the fun out of skiing; in fact, we used our skis rather as a kind of glorified snowshoes. For steady and reliable travelling up and down the hills, and for hauling loads, we had seal skin under our skis permanently. Further, since we had to travel long distances over the fjord ice, which had to be probed now and then, we preferred a strong ice-chisel to the conventional sticks. This also gave us one hand free.

Travelling over sea ice calls for continuous vigilance. It was not difficult, however, to keep alert, since the ice formations made fascinating objects for observation and study. One formation in particular is worth mentioning. Newly formed ice often cracked in intense cold or under the stress of tidal currents. We knew these sharp cracks in otherwise sound ice from our experience on the Zuiderzee in Holland; while skating in the cold hours of the early morning one sometimes hears the ice snap, and the long thunder of these breaks, as the crack runs out over distances of a mile or more, are familiar to every skater in Holland. Some such cracks may be a foot wide. Immediately after their formation they begin to freeze over; ice needles grow from the two opposite edges, and soon the lane is covered by a new black pane of clear ice, with a seam in the centre where the growing edges met.

In Greenland we sometimes found complicated cracks of this kind, running across the mouth of a fjord. Cold, or the outgoing tide, would tear two icefields a foot or two apart. When the next tide came in, the outer field would be pressed back against the inner field and the thin ice covering the crack was crushed. Its débris, lifted up vertically, froze as a winding, frayed ribbon in the crack. The next outgoing tide would then tear the two icefields apart again, and now two lanes would open, one on each side of the irregular ribbon which stayed more or less in the centre. These two narrower lanes would again freeze over, each forming a seam in its centre. In this way a system of parallel ice ridges formed which ran the full length across the mouth of the fjord. In my photograph the primary and two secondary ridges are shown, and the seam in the left 'tertiary' lane, which was to form one of the four tertiary ridges, is also visible.

Animal life on land was scarce in this season. Among the higher animals the only vegetarian that kept going, in a quiet, unassuming way, was the Ptarmigan. Like our Blackcock it lived entirely on the leafy parts of plants containing much cellulose such as twigs—not berries—of *Empetrum*, the Crowberry. The best patches of this plant grew in places that were usually covered with heavy snow, and the birds spent part of their day digging underneath it. While skiing over land we sometimes felt movement underfoot, or saw the snow move under the ski of the person ahead, and all at once a frightened Ptarmigan would burst out of the snow and rocket away. They were at this time in their lovely white winter plumage and often merged completely with their surroundings—except for their dark eyes. In flight the black outer tail feathers were shown, and these no doubt acted as a follow-my-leader signal in the flock.

Vegetarian mammals were absent, but the Arctic Fox was common; in winter it preyed mainly on the Ptarmigan. Fox tracks ran all over the country; like our Fox the Arctic species covers large areas on its hunting trips. We saw more of the Foxes in summer than in winter, probably because in summer we spent more time just sitting and watching. Once in August we came upon a mother with two cubs, absorbed in playing with a large half eaten Char.

Ravens were not uncommon. They lived on a very varied diet. Many, we were told by the Eskimos, followed the Polar Bears on their wanderings over the outer pack ice and lived on the remains of the seals the bears killed. We often saw them busy pecking at the ice; again the Eskimos told us that they could reach dead fish through the ice, but we never saw them do this. But what we did see ourselves was even more astonishing: they chased Ptarmigans on the wing and sometimes obviously succeeded in killing them. They also finished the meals of the Greenland Falcons. These beautiful birds were not at all rare, particularly in autumn. They are larger than Ptarmigan and many were of the magnificent white variety, dotted with small black specks, which made their plumage look like a coat of ermine. In their flight they were different from Peregrines, being more leisurely. Under certain conditions we have confused them with gulls. Most of them preyed on the Guillemots living in the outer fringe of the pack ice, and when the pack closed in on the land they often extended their hunting grounds and tried their hand at the Ptarmigans as well. We never saw a Greenland Falcon in the act of taking a prey, but we often found the carcases of their victims, usually cleaned up by the Ravens and perhaps Foxes.

Of course, the conditions in the sea, where the temperature remains mild, were never so extreme as on the land, and life in the sea was rich. Most of the life on land in these regions is ultimately dependent on the sea. This was certainly true for Man. The Eskimos lived on the harvest of the sea, chiefly seals. The Bladdernose, the Greenland Seal and the Bearded Seal all left for the open pack when the fjords froze over, but one species, the Fjord Seal, remained. Large numbers, many of them coming from breeding grounds further north, settled in the fjords of our district for the winter, where they lived under the ice. With their strong teeth they gnawed holes in the ice through which they breathed. This is the seal that is caught by patient Eskimo hunters sitting and watching at their breathing holes and harpooning them when they come up.

We could make a rough assessment of their numbers in April, when on warm days they crawled on to the ice to bask in the sun. In one fjord, Qingorssuaq, which was about half a mile wide and 12 miles long, we counted about 160 seals per mile. This was after a full season's hunting by several men from Kûngmiut. Since almost all these seals were offspring from stock that bred in uninhabited fjords further north, this seems to indicate that human predation in the area as a whole was negligible. With the Bladdernose Seal it is a different story; living in the open pack ice they are within reach of European sealers.

The Angmagssalingmiut applied various methods to catch Fjord Seals. In open water they shot them from the shore or from the ice; in summer, they harpooned them from kayaks. When the seals lived under the ice, they could be caught in nets or harpooned when they came up to breathe; or a 'long harpoon' was used to spear them under water through a breathing hole. This was done by two men: one lying flat on the ice, peering down a breathing hole, and steering the harpoon; the other standing by, attracting the seals' attention by scraping the ice with his ice-chisel and, at the same time, holding the harpoon, ready to thrust it down when the observer, having aimed carefully, initiated the move.

When a seal was hit, the barbed point of the harpoon anchored itself firmly in its body and came off; the hunters then hauled the seal up by a long line attached to the point, while the harpoon shaft was pulled up separately. When used skilfully this could be an effective method; once we saw two men kill five seals in three hours.

When we had been living in Kûngmiut for about six weeks and had been accepted as members of the little community, we began to build up the collection of typical Eskimo equipment which we had been asked to acquire for a museum in The Hague. We let it be known among all Angmagssalingmiut that people in Holland had heard of their technical skill and were anxious to see proof of this themselves. We also spread word that we had been given sugar, tobacco, beads, fabrics and many other desirable things to trade. We further explained that what we wanted most were objects that had actually been used, but that, since these things would be exhibited in a place where thousands of *Pukitsongmiut* ('Lowlanders') would come to admire them, we were interested only in the best specimens of their craft—we would hate to give our countrymen the impression that the Eskimos were incompetent in any way.

The response to our message was good. Once our tobacco and sugar had been sampled, it became overwhelming and we soon had to be very critical and selective in our dealings. Once, in a remote village, during a meeting specially called for the purpose, one hunter surprised his friends by offering us some very handsome, though a little battered, household goods. When questioned by his friends—he was a good hunter who was never short of anything he could get in the Government shop in exchange for his sealskins—he said, apologetically: 'Oh, my wife wants me to make some new stuff anyway.'

Thus, all through the winter, life was never dull. Yet we were looking forward to the spring, when we could start on our main work—a study of the breeding behaviour of some Arctic birds. For reasons to be explained later, our main interest was in the Snow Bunting, a small song-bird which was abundant here, but which was a migrant and did not arrive until about the middle of March. By the end of February we had finished most of our collecting, we had thought out various alternative plans in every detail, and had checked, repaired and added to our equipment.

March came, and the winter was more grim than ever. Temperatures were lower than before and the snow still piled up during occasional spells of stormy and cloudy weather. But the nights were already shortening and the long and often sunny days were ideal for outdoor work. We spent our time visiting small Eskimo communities in other fjords, completing our collections and making long hunting trips: all the time continuing our general observations.

One day, March 22nd, we stayed at home, finishing our large Char-net that was going to provide us with fresh food later in the summer. A furious *Neqajâq* (blizzard) was raging outside, but it came from the east and the temperature, though still below freezing point, was unusually mild. Suddenly the door was flung open and a breathless youth tumbled in, calling excitedly: 'The Snow Buntings are coming, the Snow Buntings are coming!' We rushed out and there, huddled behind a stone, sheltering from the blinding snow, were three tiny birds. They were Snow Buntings all right. The excitement among the Eskimos was great. The winter was over! Though how on earth those small birds would survive in this bare, snow-covered land we had not the slightest idea.

This seemingly trivial event changed our life entirely. This was what we had been preparing for. We were fully determined to make the most that the short spring allowed us for observations which we might never be able to repeat.

CHAPTER THREE

SNOW BUNTINGS
AND PHALAROPES

The arrival of the first Snow Bunting meant so much to us because we had selected this species for our spring work. A common, conspicuous bird living in a very open habitat and therefore easily observable, a close relative of the European Buntings that Eliot Howard had studied so closely, it seemed to be an ideal object for a detailed field study. Further, there were reports in the bird litera- ture indicating absence of a territorial system in this species, and since its nearest relatives were typically territorial birds, a comparison might give us some clue to the significance of territory in birds. As it turned out, the Snow Bunting was as territorial as any other Bunting, but that was not serious; a territorial bird was just as interesting to us and we were confident that the study of any bird under such favourable circumstances was worth while.

From now on we had to be up with the birds, which meant rising about two hours before sunrise so as to be able to start at first light, which of course in these re- gions is longer before sunrise than it is at our latitude. With the approach of the Arctic summer we had to get up earlier and earlier, and from the end of April on we had to wake up at midnight. Our daily rhythm thus be- came peculiar: field observations from 1 a.m. until about 9 or 10; then back to our camp, for a hearty breakfast;

then a couple of hours' working out the day's observations; lunch, an hour's sleep, some more time for observations, a good meal, and by 8 or 9 we turned in for another 3 to 4 hours' sleep. Although it is quite amazing
how the Arctic summer reduces one's demand for sleep,
we did not manage to get enough of it and it often took
us quite an effort to obey the alarm clock. One of those
days I did one of the strangest and silliest things in my
life. I used to stop the alarm clock at its very first ring,
by seizing it with one hand and turning it off with the
other. I normally lay on my right side, facing the clock,
but one morning I happened to have turned on my other
side, facing my wife. When the clock rang, I made the
usual movement, still half asleep. I became annoyed
when the clock refused to be silenced. I pressed harder,
but nothing happened, and suddenly, opening my sleepy
eyes, I realized the idiotic truth: I had my poor wife's
head in a firm grip and was pressing the thumb of my
other hand against her nose, exerting more and more
pressure to stop that wretched clock. I have rarely felt
so silly. However, it woke us up—especially my wife.

The first flocks of Snow Buntings were small. They
lived exclusively round the human habitations and on
the ruins of old but now abandoned Eskimo houses.
These were the places where, thanks to the manuring by
dogs and men, the vegetation was more luxurious than
anywhere else. Thick carpets of grass often protruded
above the snow, their seeds providing good food for the
birds. This was a fine demonstration of the general fact
that, in the Arctic, much of the life on land is ultimately
dependent on the sea, for it is the sea that feeds both
dog and man.

New waves of migrant Snow Buntings came with each
warm and humid easterly gale, usually during heavy
snowfall. In the course of April the flocks roaming over
the country became larger. In the first few weeks we saw

exclusively males; as in so many other migrant species, the females arrived later.

At first the birds were completely social, always moving about together. During this period we saw some interesting 'natural experiments,' showing the function of their call note, a long-drawn 'pee.' It sometimes happened that one or two birds stayed behind when a flock took wing and such birds sometimes called the 'pee' note. In such cases the flying birds turned at once and alighted again.

In the middle of April the behaviour of some of the birds began to change. They called more and more often and some even began to utter short, subdued phrases of song. Now and then a male would stop foraging and run at another male in a flat horizontal posture, its head withdrawn between the shoulders. The other sometimes turned to face such a charge and for just a moment the two would flutter up against each other. Such males were about to leave the flock: a few days later they went off on their own and each took up a territory.

We saw the first territorial male on April 21st, a full month after the first birds had arrived. Soon more males settled, at first far apart. The localities they chose looked very different from what they would be like in the breeding season; they were still covered with a thick blanket of snow many feet deep, only the tops of a few large boulders being visible. It was quite amazing how the birds could know whether they were on suitable territory and even on land at all; some flat stretches did not really look any different from the fjord ice. Yet none ever settled on the ice. Most males established their headquarters on some high boulder; and here they started to sing. Their song had the typical timbre of all the Buntings; it was simple, but varied considerably from one male to another. After the grim winter, when the silence could hang almost tangibly over the valleys, the revival first of the birds' song, then of the soft murmur of the brooks, delighted us beyond description.

The singing males were continuously on the alert and every now and then they flew up and rose high into the air to perform a sort of song-flight, then sailed to another boulder to resume their song there. Whenever another Snow Bunting came near, the resident male adopted the forward posture we had already seen in the birds in the flock, called a sharp 'pee' and flew to the attack. Usually this made the intruders leave at once. I often watched such strangers closely. Sometimes they belonged to flocks that were quietly foraging and just happened to stray into occupied territory. When such birds heard the song of the owner, they showed signs of intense fear; sleeking all their feathers, they crouched on the ground, looking round intently. Once they saw the other bird, they followed him with their eyes, ready to bolt at the first sign of attack. Several times I could be sure that such a bird had not yet seen the other when it crouched and it must have responded to the song only.

Some intruders were themselves in possession of a territory, and they were not always ready to flee, but might put up a defence. In such cases we saw prolonged fights. The most spectacular encounters were the 'pendulum flights', when each male alternated with clocklike regularity between attack and escape, each pursuing when the other turned tail. Like two giant black-and-white butterflies the two, keeping distance as if connected by an invisible rod, fluttered back and forth over the snow. It was one of the most beautiful things I have ever seen.

It was extremely interesting to follow, from day to day, what happened in the area we had under observation. At first the males were few and wide apart and defence of their territories was intermittent and not very fierce, except near their song posts. But as more birds settled, fighting grew more intense. Usually the original owners had to give way to newcomers and yield part of their original ground to them, but later we saw several newcomers trying in vain to get a foothold. It was not at all rare to see a male fly round and settle now here, now there, and be chased away whenever he alighted. But such males, while obviously no longer attached to the flock, and searching for a territory, were not persistent; they might not yet have settled even if there had been no resistance. Some males, however, did persevere, and became involved in grand fights. Yet of these serious competitors several were unsuccessful and even after two days of continuous fighting, in which they seemed to succeed at first, they might still be thrown out.

These observations made a deep impression on me, and ever since I have been convinced that territorial fighting actually had the function of spacing-out the breeding pairs—an opinion that subsequent observations have been unable to shake. Yet it is true that this kind of observation cannot really prove the point; and some consistent experimenting on this might dispel what doubts there still are.

The variation in the males' songs was striking. The interesting thing about it was that there were often little areas where three or four neighbouring males sang roughly the same song—where a local 'dialect' was spoken. We found also that similar dialects existed at localities a mile or more apart.

Such dialects could, of course, mean one of two things: either the males living near together, and singing the same dialect, were genetically similar—perhaps father and sons and grandsons—or they had learned from each other. Field observations could not decide this, but since then the experimental work done by Dr W. H. Thorpe in Cambridge has shown that Chaffinches and various other song-birds learn several aspects of their song by imitation, and it was possible to induce artificial dialects in birds by allowing them in their first spring (which is the critical period for this learning process) to hear abnormal song.

The boundary fights were extremely interesting to watch. Two males engaged in a 'pendulum flight' would sometimes come to grips. They held each other with their claws, pecking furiously and flapping their wings. Often they fell on the snow and sometimes rolled down hill tumbling over and over, forming one confused tangle of black and white. When they separated they faced each other in the forward posture and often, much to our surprise, made sudden pecking movements at the snow, as if picking something up. We did our utmost to find out what they could peck at, but finally came to the conclusion that they did not take up anything at all; often the bill did not even touch the ground. This was the first time I became aware of a phenomenon that was later called 'displacement activity' and it puzzled me a great deal. It still does, although now something more is known about them than at that time. The occurrence of such 'irrelevant' movements as parts of displays of various kinds was known and had been described, for instance by Selous and Huxley, but what caused them to appear only during fights and in courtship was not then fully understood.

Another interesting aspect of the males' behaviour in this period was the influence of the weather on song and other territorial activities. Snowfall always stopped it abruptly. At first we thought that this was because the snow covered the feeding areas, which would make it necessary to get down at once to the business of feeding. This may well be the general function, but the birds did not respond to the feeding grounds being covered, for slight snowfall on a warm day would stop them, although the snow would at once melt on the exposed soil.

On the other hand, when the sky cleared after a heavy snowfall, and the new snow-carpet was really covering much of the ground, the males sang with full vigour although it was still cold. Obviously, they themselves responded automatically to the stimulus 'snow falling', which, as a rule, of course, was a sign of deteriorating feeding conditions. When song stopped, it was not long before the males abandoned their territories and joined the flocks again. As in other birds, there was a regular daily rhythm, the birds rejoining the flocks at the end of each morning; and this return to social behaviour occurred later and later in the day as the season advanced. The weather superimposed 'corrections' on this basic rhythm and thus the actual moment of the switch-back fluctuated considerably from one day to another.

Numerous other little things happened that kept us busy noticing and recording. As the season advanced, we saw more and more details. For instance, the song activity of each male was always correlated with increased alertness. A male starting to sing would always hop or fly to an elevated lookout post; here it would not only watch but also listen, and as soon as it heard the call note of another Snow Bunting it would stretch its legs, body and neck and look round. We could make them do this by imitating the call note ourselves. Their failure to locate the call, however, was as striking as their extreme sensitivity, for they heard the slightest Snow Bunting sound, but never seemed to know where it came from.

All these things were happening in the absence of females. We did not see the first female until the end of April, more than a month after the arrival of the first males.

When the females began to arrive, the picture became much more involved. At first the females joined the birds still living in the flocks. We kept a careful watch for

Fig. 2. Male Snow Bunting preening.

females straying over occupied territories, for we were keen to see the complete courtship of the males. To our surprise, the first male we saw respond to a female flying over his territory assumed the forward posture and gave the call that we had heard in hostile encounters between males! We thought at first that this was due to the great distance and that the male just could not have seen that he had to do with a female, but it was soon obvious that all males behaved in this way every time they saw a female, even when she appeared right in front of him and quite near. Remembering all the proofs of extremely acute vision we had seen in our birds (such as looking up at flying birds visible to us only with the aid of binoculars), we could not believe that the male's eyes could not distinguish between the sexes. At the time, I merely reported this as a fact: the male's initial response to females is just the same as that to males, and it seems to be a hostile response. The only thing I pointed out at the time was that the male did show discrimination in his later behaviour; when facing a female he follows up by courting her, whereas a male is attacked. This showed that the male's behaviour is a chain of successive acts.

Since then our knowledge of bird displays has grown considerably and we know now, for instance, that a hostile initial response to females is the rule in many territorial species. This seems to be due to the fact that the female provides in part the same stimuli as an intruding male; and the males of such species simply must respond aggressively to other males because this is the response on which the spacing-out of breeding pairs is based. This again is now seen, perhaps more clearly than 25 years ago, to be an absolutely vital process. I will discuss present views of this whole complex of phenomena in some more detail in Chapter Twelve.

At the time the full significance of this curious threat behaviour of the males in front of females was not clear to me. It is realised now that the adoption of a threat posture is due to more than a mere tendency to attack; it is due to an internal conflict in the bird between attacking and fleeing—we must assume that such a male is angered by and afraid of the female at the same time. The fear component is also suggested by the fact that the courtship of the male consists in part in running away from the female. During these short runs he spreads his wings and tail, and so displays his wonderful black-and-white plumage to full advantage. After this he quickly returns to the female, often in the threat posture and then runs away again, displaying. In view of recent and more precise work on the behaviour of the related finches it seems probable that this curious courtship of the Snow Bunting may have originated in a real fear response of the male.

It is not my intention to describe here the many interesting details of the life history of the Snow Buntings that we saw. I have recalled some of our observations because, in retrospect, it is clear that they were the starting points of several later studies; conversely, the results of these later studies made some Snow Bunting facts more intelligible.

During May the winter changed almost imperceptibly into spring. The mixed chorus of Snow Buntings, Lapland Buntings, Wheatears and Ptarmigan, to which the mountain brooks added their babble, grew louder every day, until finally even the nights were not cold enough to silence the murmur of the water completely. Yet the snow, wet and soggy by day, froze solid every night, and walking to our observation posts became quite unpleasant. We needed our skis for crossing the fjord ice, which by now began to grow treacherous in places, and for keeping on top of the slushy snow later in the day. On our way out we were often tempted to leave our skis behind as soon as we had crossed the fjord, which was all right as long as we were back at the place we left them before the snow began to turn soft. All went well a couple of times, but we misjudged the situation one warm day and found ourselves stuck in soft slush, still a mile away from the skis. That morning taught us a valuable lesson. Sinking waistdeep into the snow at every step, we found that our only hope was in crawling, and this was how we decided to proceed. Maybe this sounds neither funny nor particularly unpleasant to one who has not been in East Greenland. In reality, in that country of tremendous snowfall it is both. We were extremely relieved when we finally reached our skis again.

Owing to the ever-changing conditions of ice and snow we had to move to new observation areas a couple of times. This naturally interrupted our records of individual birds, but we could make up for this handicap by catching up at each move with birds that were less advanced than those we had just left.

In June, the fjord ice broke up and we moved out altogether and established a camp on a small island in the middle of the fjord, where we were out of reach of the Huskies, which were by now scavenging all along the seashore. On this island we lived a delightfully simple life for several months. Almost all our time, nearly 20 hours a day, was available for our work. As food we had rice with some dried vegetables, both used sparingly and as a little addition to our main food—fresh Char from the fjord. Not more than half an hour each day was needed to paddle across the fjord and back, which included a short visit to our fish-net.

The Char was quite an interesting fish apart from being a delicious food. We saw some striking examples of differences between local populations. All Char spent the winter in large fresh water lakes. When the ice in a fjord broke up, the fish went to sea. Each fjord had its own type of Char, different from others in size, colour of flesh, skin colour, etc. What we saw of the stomach contents proved that the food of the different populations was equally different: in one fjord they lived mainly on small fish, in our own fjord, which was a 4 miles long side-fjord of this other fjord, the food was mainly crustaceans. This might very well account, at least in part, for the differences in appearance.

Our own fjord, Torssukátaq, was separated from another fjord (Tunoq) by a low ridge, which was covered only by spring-tides. The local Eskimos had already told us that the first spring-tide after the break-up of the ice in Tunoq would carry Tunoq Char into our fjord. The ice in Tunoq broke weeks later than that in Torssukátaq and, true enough, the day after the next spring-tide we caught two strange Char in our net, which we found later were similar to Char found in Tunoq.

Of the numerous interesting experiences we had during this wonderful summer I will discuss only one more in some detail—our meeting with Red-necked Phalaropes, 'the Amazons of the bird world.'

Phalaropes are small waders living in the arctic and subarctic zone. They are one of the few groups of birds in which the females are much more brightly coloured than the males. It had been known for a long time that the males, which have the same kind of camouflaged plumage as, for instance, Snipe, do all the incubating and guard the young alone. The females just lay the eggs and have none of the traditional motherly chores.

We were aware of the possibility that we might see these birds on their breeding grounds and that, with luck, we might see something of their courtship. If it was true that the brighter coloration of so many male birds was connected with their display movements, making them more conspicuous to members of their own species, then a species with reversed sexual dimorphism must also have 'reversed' courtship, or, in general, display—i.e., the female must do the displaying instead of the male.

When, therefore, at the end of June we happened to see some Phalaropes in a little pool near our Snow Bunting area, we decided at once to cut down a little on our Snow Bunting programme and instead to give some attention to the Phalaropes.

These first birds we saw were all females and were extremely tame. Their back was a beautiful grey, with two parallel yellow lines along it. The sides of the neck were a warm chestnut brown, fading towards the head into dark grey. The chin was white and the whole underside very light, with grey mottling.

The females, we soon saw, isolated themselves in the way the male Snow Buntings had done; but the habitat they chose was the marshy flats and the ponds there. Here they spent most of their day, usually feeding. They swam in the curious Phalarope fashion, floating as high on the water as a cork and, with quick, jerky movements, picking up one mosquito larva or pupa after another. Or they would 'stalk' an adult mosquito, swimming with their head flat on the water, slowly and smoothly, then making a sudden dash at the prey. Early in the morning, when the water was cold and completely calm, and the insects being motionless in the water, the Phalaropes would often do a sort of pirouette, swimming in very small circles and retracing their 'steps' in order to collect the insects that were whirled up, and so made visible, by the disturbance of the water.

We observed those isolated females for several hours each day. Now and then one of the females suddenly stopped feeding, flew up with a curious loud buzzing of the wings and, calling a harsh, staccato 'whit—whit—whit', flew some ten or twenty yards low over the surface of the water and, alighting again, called something like 'wedu—wedu—wedu', swimming with its neck stretched and looking round intently. At first we saw this ceremony on rare occasions only, but soon each female began to do this frequently, every five minutes or so, for hours on end.

We saw later that this ceremony attracted males and that other females tended to avoid the neighbourhood of such displaying females. The males, which arrived soon after the females, never performed in this way and there seemed little doubt that the ceremony was the song of the species, and that in Phalaropes song is an attribute of the female.

In another respect, too, she assumed a rôle usually played by males; she vigorously attacked rivals—i.e., other females visiting her pond—and all the fights we saw were between females. Visiting females did not always flee; sometimes they just evaded the attacks without leaving the pond and after a while they were tolerated, at least as long as they did not display.

Males visited the females on their ponds. As with the Snow Buntings, we were keen on seeing the first encounters between the sexes. The first response of a female to a visiting male was striking, though very much in line with what we had seen in the Snow Bunting: she flew towards him, often uttering her song and, after alighting, faced him in the same flat, threatening posture she adopted when attacking another female. But she never drove her attack home; a foot or so from the male she suddenly stretched her neck and swam away. Often she alternated several times between approach in the forward threat posture and withdrawal with raised head. Again, in this species, the first response to a bird of the other sex is the same as that to one of the same sex, and the posture shown is a hostile one. Also, there are traces of fear or at least avoidance in the subsequent behaviour. Although I described this behaviour at the time, I did not realise its full significance until much later.

The particular female we happened to concentrate on remained unmated for a long time, a fact that gave us another very convenient 'natural experiment'. She began to sing more and more frequently and, at the same time, began to respond to birds other than Phalaropes and flew towards them in the same way as females generally did to male Phalaropes. The birds found round her pool were, in order of frequency, Snow Buntings, Lapland Buntings, Ringed Plovers and Purple Sandpipers. Of these, all except the Snow Buntings drew such 'erroneous' responses. Two things interested us about this: (1) these three species looked rather like a Phalarope, being about the same size and general colour, whereas the Snow Bunting had a conspicuous white wing patch, and (2) the erroneous responses grew more frequent later in the cycle when the female also sang much more often than before. It was obvious that, as the female's reproductive motivation grew, she began to be less selective in her response. When, after such a mistake, she alighted near the strange bird, she stopped and, after some hesitation, lost interest.

These observations were no more than a mere beginning; they did not allow me to draw hard and fast conclusions, but they were part of a series which helped me to form certain ideas about the organisation of such behaviour which I have found very helpful in later studies.

The interpretation of the female Phalarope's song as a ceremony which attracted males was strikingly confirmed a few days later. Our female acquired a mate in due course; the birds lived and foraged together, and copulated several times. Now and then the female went on land and made a 'scrape' by sitting down and scrabbling backward with the feet in the same way as other waders do. Usually the male joined her and soon the birds had made a number of scrapes at various places near the pond. Some of them they visited more than once, others seemed to be forgotten soon after they had been made.

One day, while the two birds were quietly foraging on the pool, the female suddenly uttered her 'song'—which we had not heard since she had acquired a mate—and flew up. But instead of landing on the water, she flew into the marsh and went to a scrape. The male followed her at once, which he did not usually do when she flew off, and together they busied themselves there for a couple of minutes. All at once the female sang again, flew off once more, again taking the male with her. She now went to another scrape. And in this way the pair went from one scrape to another (incidentally showing how well at least the female knew where they were). In the fourth scrape the female stayed a little longer and there she laid her first egg.

I was particularly struck by this revival of the song and by the male's prompt response to it. I missed the laying of the second egg because a gale prevented me from crossing the fjord. The next day, expecting the bird to lay her third egg, I took no chances and kept a continuous watch on the pair from 1.15 a.m. I expected to see her lay her third egg some time early that morning. As it turned out I had to keep up my vigil until 5 p.m., when the birds rewarded me by going through exactly the same ceremony again.

I was so pleased to see this because it confirmed, in a way, the general biological 'philosophy' which is behind so much of a naturalist's watching. I understood that this bird *had* to have something like this ceremony and that the male *had* to follow the female so promptly; because she lays the eggs, whereas he has to sit on them, there must be some way in which either he learns where to incubate or (if the male selects the nest site) she learns where to lay her eggs. Obviously the 'nest-showing' ceremony was the means by which the female made the male follow her to where she would lay the eggs, so that he could learn their whereabouts.

Such nest-showing ceremonies occur in many animals. In some birds—e.g., many hole-breeders—the male selects the nest site and points it out to any female that shows interest; this has been described, for instance, in Redstarts, Pied Flycatchers and Kestrels. In sticklebacks, which I studied later, I saw a wonderful parallel to this: here, as in the Phalaropes, the male guards the eggs, but unlike the Phalarope, he builds the nest alone and it is the male which guides the female to the nest.

Our studies of the Phalaropes concluded our fourteen months' stay in 'the Friendly Arctic' and not long afterwards we reluctantly left Greenland's fjords. Faced soon afterwards with the choice between returning to Greenland or starting systematic studies of animal behaviour nearer home, I chose the latter course, but not without a difficult struggle, and neither my wive nor I have been able to forget entirely our longing for the Arctic.

Twenty-three years later, in the summer of 1956, I had two experiences which revived all our happy memories of Angmagssalik: in July, together with friends in California, I happened to visit Mono Lake in the Sierras not far from Yosemite Park and there, pirouetting and hunting the countless flies basking on the pebbles along the shore, were many tens of thousands of Phalaropes—most of them our Red-necked species or, as they are called in the States, Northern Phalaropes. All my Greenland memories came to life again.

Two days later I was carried, in embarrassing luxury, high across Greenland in a comfortable aircraft and, to my delight, our route led us exactly over Angmagssalik. As we approached the east coast from the west, the weather cleared while we were still over the vast ice cap. Soon I saw the first *nunataqs*, then the whole rugged coastal mountain range. Deep below I saw Sermilik fjord, covered almost entirely with ice and bergs, Angmagssalik fjord, Torssukátaq, Sermiligâq, the white specks of pack ice and icebergs floating in the blue-green sea, local fog veils hanging in quiet corners of the immense fjords. Ten minutes of living through our Greenland days again—then we headed east and the mountains faded into the haze behind us.

CHAPTER FOUR

BACK TO THE BEE-HUNTERS

During the return voyage from our fourteen months' stay in the Arctic, my thoughts often went back to *Philanthus*. I felt that my results obtained so far were only a beginning and I kept pondering about possible ways of penetrating more deeply into the problems of their homing. The chance to do more field work came when my professor, Dr H. Boschma, seeing that not much could be achieved in the official fourteen days' annual vacation to which a demonstrator was entitled, allowed me to take some undergraduates with me, accepting the work they would do with me as part of their training.

My boasts about the wasps' achievements had stirred some of my friends in the lab., and consequently they were only too keen to join me. The observations to be described in this chapter were done with their enthusiastic help. Our team consisted, apart from myself, of W. Kruyt, D. J. Kuenen, R. J. van der Linde and G. van Beusekom.

We began by investigating the wasp's 'locality study' a little more closely. As I mentioned before, we had already quite suggestive indications that it really deserved this name, but clear-cut proof was still lacking. The otherwise annoying vagaries of the Atlantic climate provided us with a wonderful opportunity to get this proof. Long spells of cold rainy weather are not uncommon in a Dutch summer—in fact they are more common than periods of sunny weather, which alone could tempt the wasps to 'work'. Rainy weather put a strain on morale in our camp, but the first sign of improvement usually started an outburst of feverish activity, all of us doing our utmost to be ready for the wasps before they could resume their flights.

We had previously noticed that many (though not all) wasps spent cold and wet periods in their burrows. Rain and wind often played havoc with their landmarks and perhaps the wasps also forgot their exact position while sitting indoors. At any rate, with the return of good weather, all the wasps made prolonged 'locality studies' when setting out on their first trip. Could it be that they had to learn anew the lie of the land?

On one such morning, while the ground was still wet but the weather sunny and promising, we were at the colony at 7.30 a.m. Each of us took up a position near a group of nests and watched for the first signs of emerging wasps. We had not to wait long before we saw the sand covering one of the entrances move—a sure sign of a wasp trying to make her way into the open. Quickly we put a circle of Pine cones round the burrow. When the wasp came out, she started digging and working at her nest, then raked sand over the entrance and left. In the course of the morning many wasps emerged and each received Pine cones round her entrance before she had 'opened the door'. Some of these wasps did not bother to work at the nest, but left at once after coming out. These latter wasps we were going to use for our tests. As expected, they made elaborate locality studies, describing many loops of increasing range and altitude before finally departing. We timed these flights carefully. As soon as one of these wasps had definitely gone, we took the Pine cones away. This was done in order to make absolutely sure that, if the wasp should return

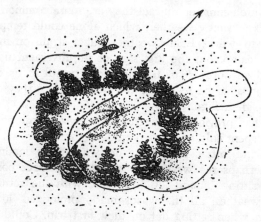

Fig. 3. Philanthus making a locality study.

unobserved, she could not see cones round her nest. If then, when we saw her return with a bee, a displacement test in which the circle of Pine cones was laid out some distance away from the nest would give positive results (i.e., the wasp would choose these cones), we would have proved that she must have learnt them during her locality study, for at no other time could she have seen them.

Not all such wasps returned on the same day. Their prolonged stay and their fast down in the burrows probably forced them to feed themselves in the Heather first. Some, however, returned with a bee and with these we succeeded in doing some exciting tests. In all we tested 13 wasps. They were observed to choose 93 times between the true nest and a 'sham nest' surrounded by the Pine cones. Seventy-three choices fell on the sham nests, against only 20 on the real nests. In control tests taken after the experiments, when the cones were put back round the real nest, of a total of 39 only 3 choices were now in favour of the sham nests, the other 36 being in favour of the real nests. There was no doubt then that these wasps had learnt the nature and the position of the new landmarks during the locality study.

The most impressive achievement was that of wasp No. 179. She had made one locality study of a mere six seconds and had left without returning, let alone alighting. When she was tested upon her return more than an hour later she chose the cones 12 times and never came near the nest. When the original situation was restored she alighted at once on her burrow and slipped in. Nos. 174 and 177 almost equalled this record; both were perfectly trained after uninterrupted locality studies of 13 seconds. All the other wasps either made longer locality studies or interrupted them by alighting on the nest one or more times before leaving again. Such wasps might have learnt during alighting rather than while performing the locality study, so their results were less convincing.

This result, while not at all unexpected, nevertheless impressed us very much. It not only revealed an amazing capacity in these little insects to learn so quickly, but we were struck even more by the fact that a wasp, when not fully oriented, would set out to perform such a locality study, as if it knew what the effect of this specialized type of behaviour would be.

I have already described that a wasp, which has made a number of flights to and from a burrow, makes no, or almost no, locality study, but that it will make an elaborate one after the surroundings have been disturbed. Further tests threw light on the question what exactly made her do this. We studied the effect on locality studies of two types of disturbances. In tests of type A we either added or removed a conspicuous landmark before the wasp returned and then restored the original situation while she was inside. Such wasps, although finding the old, familiar situation upon emerging again, made long locality studies. In tests of type B the wasps were not disturbed at all when entering, but changes similar to those of the A-tests were made just before they left. None of these wasps made locality studies. Wasps used for A-tests always hesitated before alighting. Therefore, disturbances of the familiar surroundings perceived upon returning make the wasps perform a locality study when next departing, while the same disturbances actually present at the time of departure have no influence!

Some further, rather incomplete and preliminary tests pointed to another interesting aspect. Conspicuous new landmarks given before the return of the wasp and left standing until after her departure influenced the form of the locality study as well as its duration: the wasp would repeatedly circle round this particular landmark. If, however, such a landmark was left for some time, so that the wasp passed it several times on her way out and back, and then moved to a new place, the wasp would make a longer locality study than before, yet she would not describe extra loops round the beacon. She obviously recognized the object and had merely to learn its new position. These tests were too few and not fully conclusive, but they did suggest that there is more to this locality study than we had at first suspected. The whole phenomenon is remarkable and certainly deserves further study.

We next turned our attention to the exact nature of the landmarks that were used by the wasps. What exactly did they learn? We spent several seasons examining this and the more striking of our tests are worth describing.

First of all we found that not all objects round the nest were of equal value to the wasps. The first indication of this was found when we tried to train them to use sheets of coloured paper about 3 × 4 inches, which we put out near the nests, as a preparation to study colour vision. It proved to be almost impossible to make the wasps use even a set of three of them; even after leaving them out for days on end we rarely succeeded with the same simple displacement tests that worked so well with the Pine cones. Most wasps just ignored them. Yet the bright blue, yellow and red papers were very conspicuous to us. For some reason, the Pine cones were meeting the wasps' requirements for landmarks better than the flat sheets.

Together with Kruyt I worked out a method to test this. We provided two types of objects round a nest —for instance, flat discs and Pine cones—arranged in a circle in alternation. After a day or so, we moved the whole circle and checked whether the wasps used it. If so, we then provided two sham nests at equal distances, one on each side of the real nest, and put all objects of one type round one of these sham nests, all of the other type round the other. If then the wasp had trained her-

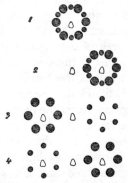

Fig. 4. 'Preference tests':

1. *The training situation*   3. *The two kinds of landmarks*
2. *First check*   *separated*
                                           4. *Reversal of 3*

self to one type of landmark rather than to the other, it should prefer one of the two sham nests. Such a preferential choice could not be due to anything but the difference in the wasps' attitude towards the two classes of objects, for all could have been seen by the wasp equally often, their distance to the nest entrance had been the same, they had been offered all round the nest, etc.—in short, they had had absolutely equal chances.

In this way we compared flat objects with solid, dark with light, those contrasting with the colour of the background with those matching it, larger with smaller, nearer with more distant ones, and so on. Each test had, of course, to be done with many wasps and each wasp had to make a number of choices for us to be sure that there was consistency in her preference. This programme kept us busy for a long time, but the results were worth the trouble. The wasps actually showed for landmarks a preference which was different from ours.

When we offered flat circular discs and hemispheres of the same diameter, the wasps always followed the hemispheres (43 against 2 choices). This was not due to the larger surface area of the hemispheres, for when we did similar tests with flat discs of much larger size (of 10 cm. diameter, whereas the hemispheres had a diameter of only 4 cm.), the choices were still 73 in favour of the hemispheres against 19 for the discs.

In other tests we found out that the hemispheres were not preferred because of their shading, nor because they showed contrasts between highlights and deep blacks, nor because they were three-dimensional, but because of the fact that they stood out above the ground. The critical test for this was to offer hollow cones, half of them standing up on top of the soil on their bases, half sunk upside down into the ground. Both were three dimensional, but one extended above the ground while the others formed pits in the ground. The standing cones were almost always chosen (108 against 21).

The preference for objects that projected above the ground was one of the reasons why Pine cones were preferred. Another reason was that Pine cones offered a chequered pattern of light and dark, while yet another reason was the fact that they had a broken instead of a smooth surface—i.e., dented objects were more stimulating than smooth ones. Similar facts had been found about Honey Bees by other students and much of this has probably to do with the organization of the compound eyes of insects.

We further found that large objects were better than small objects; near objects better than the same objects further away from the nest, objects that contrasted in tone with the background better than those matching the background, objects presented during critical periods (such as at the start of digging a new nest or immediately after a rainy period) better than objects offered once a wasp had acquired a knowledge of its surroundings.

It often amazed us, when doing these tests, that the wasps frequently chose a sham nest so readily although the circle offered contained only half the objects to which they had been trained. This would not be so strange if the wasps had just ignored the weaker 'beacons', but this was not the case. If, in our original test with flat discs and hemispheres, we would offer the discs alone, the wasps, confronted with a choice between the discs and the original nest without either discs or hemispheres, often chose the discs. These, therefore, had not been entirely ignored; they were potential beacons, but were less valued than the hemispheres. Once we knew this, we found that with a little perseverance we could train the wasps to our flat coloured papers. But it took time.

The fact that the wasps accepted these circles, with

half the number of objects they used to see, suggested that they responded to the circular arrangement as a whole as well as to the properties of the individual beacons. This raised the interesting issue of 'configurational' stimuli and it seemed to offer good opportunities for experiment. This work was taken up by Van Beusekom who, in a number of ingenious tests, showed that the wasps responded to a very complicated stimulus situation indeed.

First of all, he made sure that wasps could recognize beacons such as Pine cones fairly well. He trained wasps to the usual circle of Pine cones and then gave them the choice between these and a similar arrangement of smooth blocks of Pine cone size. The wasps decided predominantly in favour of the Pine cones, which showed that they were responding to details which distinguished the two types of beacons.

He next trained a number of wasps to a circle of 16 Pine cones and subjected them to two types of tests. In type A the wasp had to choose between two sets of 16 cones, one arranged in a circle, the other in a figure of another shape, such as a square, a triangle, or an ellipse. He found that, unless the figure was very similar to the circle, the wasps could distinguish between the two figures and alighted in the circle. In those tests the individual cones did not count; he could either use the original cones for constructing the circle or use them for the square or triangle. It was the circular figure the wasps chose, not the Pine cones used during training.

In tests of type B, after the usual training to a circle of 16, he offered the 16 cones in a non-circular arrangement against 8 or even fewer cones in a (loose) circle —and found that the wasps chose the circle in spite of the smaller number of cones. He could even go further and offer a circle of quite different elements, such as square blocks (which the wasps could distinguish from

cones, as other tests had shown). If such a circle was offered against cones in a non-circular arrangement, it was
the circle that won. Thus it was shown in a variety of
ways that the wasps responded not only to the individual
beacons (as the preference tests of Kruyt and myself
had shown), but also to the circle as a whole.

However, all these experiments, while giving us valuable information about the way our wasps perceived their
environment, had one limitation in common—they
showed us only how the wasps behaved at the last stage
of their journey home. We had many indications that the
Pine cones were not seen until the wasps were within a
few yards from the nest. How did they find their way
previous to this?

Although we were aware of these limitations, it was
extremely difficult to extend our tests. However, we did
a little about this. More than once we displaced small
Pine trees growing at a distance of several yards from
nests under observation. In many cases wasps were misled by this and tried to find their nests in the correct
position in relation to the displaced tree. The precision
of their orientation to such relatively distant marks was
truly amazing.

Such large landmarks were used in a slightly different
way from the Pine cones. Firstly, they were used even
when relatively far from the nest. Secondly, they could
be moved over far greater distances than the Pine cones.
A circle of Pine cones would fail to draw the wasp with
it if it was moved over more than about 7 ft., but a Pine
tree, or even a branch of about 4 ft. high, could lure the
wasps away even if moved over 8 metres. We further
observed in many of our earlier tests that wasps, upon
finding the immediate surroundings of the nest disturbed,
flew back, circled round a Pine tree or a large sandhill
perhaps 70 yards away, and then again approached the

nest. This looked very much as though they were taking their bearings upon these larger landmarks.

Van der Linde and others also spent a great deal of time and energy in transporting individual wasps in light-proof cloth over distances up to 1,000 metres in all directions. Since good hunting grounds were to the south and south-east of the colony, whereas in other directions bare sand flats or dense Pine plantations bordered upon the *Philanthus* plains, we could assume that our wasps knew the country to the south and south-east better than in other directions—an assumption which was confirmed by the fact that our wasps always flew out in a south or south-east direction and returned with bees from there. The transported wasps, whose return to their nests was watched, did indeed much better from the south and south-east than from any other direction. From the north-west, for instance, half the wasps never returned as long as our observations lasted. This did indeed suggest that return from unknown country was difficult if not impossible and, therefore, that learning of some kind was essential, but it could not tell us more.

One test gave us a good impression of the way in which the wasps used combinations of landmarks. A wasp was given a square block of black wood, 25 × 25 cm. by 3 cm. high, near the nest and a Pine branch about 4 ft. high at a distance of about 4 ft. from the nest (Fig. 5). One of the corners of the block was touching the nest entrance. When the wasp had become used to this arrangement, Van Beusekom moved both tree and block about 10 in.; the wasp chose the block, and tried to open her burrow at *x*. When now the block was turned 45° the wasp did not know at which corner to search; it tried 5 times at corner *a*, 4 times at corner *b*, and 5 times in between.

When the tree was next moved so that it stood in position *B*1, the wasp chose 7 times corner *b*2 and 3 times

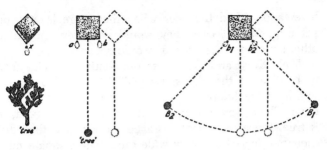

Fig. 5. Tests with block and tree.

corner b_1; when the tree was then put in position B_2, the wasp went to b_1 8 times and chose b_2 only twice. These figures are small, but the same tendency was seen in several other tests: the wasps used a variety of land-marks, some in sequence, some simultaneously.

While thus discovering, step by step, new facts about the homing abilities of *Philanthus*, we became increas-ingly interested in another aspect of their lives—their un-canny ability in recognizing Honey Bees among the multitude of insects on the Heather, many of them rather similar to Honey Bees, such as various solitary bees and, most of all, the hover fly, *Eristalis tenax*, a species that mimics Honey Bees so successfully that even birds often confuse them, as had been shown so convincingly by the experiments of Mostler. However excellent was the eyesight of *Philanthus*—as demonstrated daily in our tests on homing—their visual acuity is not very good and it was certainly impossible that they could recognize the finer details of a bee's structure. We suspected, there-fore, that a hunting wasp relied on other sense organs. Perhaps it could recognize the particular pitch of the buzzing—as we do and as some birds seem to do when they prey on drones. Or else the sense of smell might have something to do with it. Or perhaps *Philanthus* did not have to recognize bees at all, but responded to the hives and picked the bees when they were leaving or entering; in that case they would not even encounter other insects and the question would not arise.

While Kruyt and Van Beusekom continued their study of homing on the plains, I decided to spend a season watching the hunting wasps. This was easier said than done. There were a couple of thousand wasps about, it is true. But a first reconnaissance showed that they did their hunting over a very wide area, half a square mile or so. In spite of hours of watching at the nearest apiary, I never saw a wasp come anywhere near the hives, and it was therefore obvious that they did not hunt at the entrance of the hive. So I decided to tackle my problem in another way. When doing our tests on homing we had already seen that a wasp could find a bee if she had lost it (some wasps dropped their bees if we continued our experiments for too long). If we then broke off the test and put the cones back round the true nest, allowing her to find it, the wasp would open and inspect the burrow and then go back to her bee. Or a wasp might drop her prey in mid air when she was scared by a sudden movement by us. Often such wasps succeeded in finding their bee again—a remarkable achievement, since it might have been dropped yards from the nest and have landed among the dark brown moss where it was almost invisible to us.

The way in which a wasp approached a lost bee was very illuminating. She flew round in an irregular way, describing wide loops low over the ground. Suddenly she would stop, hovering in the air, and begin to make her way on a slightly zigzag course, proceeding steadily towards the bee, and finally alight an inch or so from it. Then she walked towards it, waving her antennae, and in a second grabbed it.

Now it struck us that such wasps always approached their lost bees against the wind. It was as if, on their searching flight, they were suddenly alerted by the bee's scent carried by the wind. Two simple tests showed that this was indeed the case.

When I scared a wasp, making her drop her bee, and then quickly put the bee in an open tube covered by gauze, which was dug into the ground so that the bee was invisible but its scent could escape, the wasp approached the tube against the wind in exactly the same way as when the bee was in full sight. She always ended up on the gauze, only to walk round on it, frantically trying to get in.

Another test, originally done for a different purpose, pointed the same way. As we had reasons to suspect that the wasps' organs of smell were located on her antennae —which was later confirmed—we had, in our attempts to show that homing was done by eyesight and not by scent, deprived some wasps of their antennae and subjected them to tests with landmarks. Such wasps continued working at their burrows; they flew out and made locality studies, they still followed our beacons when we displaced them—but they never brought home a single prey. This again made us think that smell played a part in hunting.

The next attempt to see something of the hunting behaviour was again made on the assumption that it would not be possible, or at least would cost an amount of time out of all proportion, to go and watch the hunting wasps in the field. I had read in Fabre's account of *Philanthus* that his wasps could be made to capture bees in captivity —in a glass jar. I repeated this, by putting a wasp with several live bees under a 3 lb jam jar. The wasp tried at first to escape, then settled down, cleaned its antennae and rubbed its legs and wings, but did not show the slightest interest in the bees. Now and then, when a bee accidentally touched her, she withdrew or sometimes turned round as if ready to defend herself. I began to doubt whether I would see anything worthwhile when suddenly the wasp, upon being touched again, seized the bee, spun round grasping it and, before I could follow what happened exactly, had stung it under the 'chin'. The bee collapsed at once, its legs twisting for a few seconds, and then stopped moving altogether. *Philanthus* now began to treat her victim in a most curious way. She pressed it tight to her body, so that nectar began to flow from its mouth. This she licked up, to the last drop. Then she moved the bee in the 'transport position' (upside down under her, held by the middle pair of legs) and flew off, only to bump into the glass and land again on the table.

I repeated this a number of times and eventually had a good picture of what happened. The wasps seized the bees only when the latter touched their antennae. Quick as lightning they turned the bee round until it was facing them and then, while the hapless bee was frantically trying to sting its attacker—in which it rarely succeeded because its sting slipped on the wasp's armour—*Philanthus*, with unfailing precision, found the chin and drove home her formidable sting.

Fig. 6. How Philanthus stings a bee.

This little drama gave me another clue. The fact that the antennae, when touched, released the wasp's hunting behaviour in its full ferocity showed that some sense organ located in the antennae was involved. Now anyone who has looked at an insect's antennae under the microscope knows that they are covered with thousands upon thousands of tiny structures of various types, and these are all connected with nerve endings leading into the large antennal nerve. These structures, which are all sense organs, have certainly a variety of functions. There are at least some which are organs of touch and of smell. I did another simple test to decide which of these was responsible in this case. Together with Honey Bees I introduced some Bluebottles and Bumblebees into the jar. Even when these insects touched the wasp's antennae they were not captured. Yet this was not decisive; they might be distinguished by touch. Now I took some small Bumblebees and shook them in a glass tube with live bees, so as to cover them with bee scent. And to my satisfacton I saw that my wasps began to show an interest in these scented Bumblebees—chemical bee-dummies so to speak—and even captured and stung one of them. This at least showed that chemical stimuli were involved—at any rate in the rather unnatural conditions of these tests.

However, I felt that these observations had not carried me very far. The wasps that retrieved lost bees might well apply a method entirely different from that applied when hunting (actually that is what I found later). And the wasps in the jam jar were not even hunting at all—they were only just willing to respond when the bee actually forced itself upon them.

So I had no choice but to go into the hunting area. But just sitting down somewhere in the vast heath and waiting for wasps to catch bees right in front of me seemed a pretty hopeless undertaking. Could I not find a part where the wasps were likely to concentrate? I reasoned that the fringe of the heath nearest to the colony must be the best area, particularly with a south-east wind when the wasps might be unwilling to fly far. Consequently I began by spending a couple of days sitting at the north-west edge of the heath and was actually lucky enough to observe the full hunting behaviour a few times. Several times a day I spotted a bright yellow wasp flying about over the heath. Some were feeding on the *Calluna* flowers, others alighted now and then on a branch or even on the ground. But sometimes a wasp flew in what seemed a searching way from plant to plant without alighting. It just hovered a few seconds, facing a bough, and then flew on to the next, to hover there. A few times I saw one of these wasps react to a bee. If she happened to come within a foot or so of one, she suddenly turned round to face it, then flew towards it and, stopping about 3 to 4 in. away, hung motionless in the air, like a hover fly. Then, with a sudden leap, she darted forward and seized the bee. And from that moment everything happened exactly as in my jam-jar observations: there was a short struggle in which the two tumbled down through the tangle of Heather, buzzing fiercely and whirling round and round; a moment later the wasp had manoeuvered the bee into position and stung it. She then pressed it and licked up the nectar, and after a minute or so she flew off, towards the north-west back to the colony.

This clearly was the complete chain! In the course of the next two days I saw several things that gave me a lead to a further understanding of it. The wasp's first response—turning round either in flight or when sitting and then facing the bee—had no relation whatsoever to the direction of the wind. Therefore, quite unlike retrieving a lost bee, this reaction could not depend on scent. I thought it must be visual, especially because the maximum distance at which a wasp seemed to discover a bee was approximately one foot, which seemed to agree well with their visual acuity, which was probably somewhere in the region of 1°. That would mean that an object the size of a bee could simply not be seen beyond about 30 cm. How, then, could they distinguish between a bee and other insects of a similar size? I soon saw that at this distance they could not. They turned and even

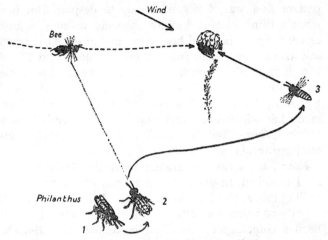

Fig. 7.   How Philanthus captures a bee.

flew towards other insects as well; I saw responses to Bumblebees, flies and small beetles. Their behaviour towards these insects gave me a further clue. They invariably approached them as they did Honey Bees, hovered near them, but then flew on. And this hovering always occurred facing the wind—whatever insect they were confronted with. It looked very much as if they were checking the scent.

Now it would not be difficult to test this experimentally, if only I could attract enough wasps to carry out some sufficiently large test series. This proved quite possible. I killed some bees and tied them to a thin thread which I hung between two twigs, placed vertically in the ground about 2 feet apart, in such a way that the bees were hanging just in front of some *Calluna* plants, and in their lee. I then sat down and awaited events. That very first day a number of wasps passed me, hunting from plant to plant. The first two went by without showing any interest whatever in my 'dummies' and, impatient as I was, I began already to despair. But just when a third wasp had approached my 'gallows', a gust of wind made the dead bees move and at once the wasp saw them, hovered in front of one of them, and threw herself on it. She stung it and, hanging upside down, one leg hooked on the thread, squeezed her victim and lapped up the nectar. Then she tried to make off—which made her whirl round and round like an acrobat on a flying trapeze. Buzzing loudly, she renewed her attempts several times, but finally she gave up.

Now I knew how to arrange my further tests. First of all I attached an extra thread to one of the twigs. By pulling this gently I could make the whole structure rock slightly and thus found that my first observation had not been a coincidence: the slight movement attracted the attention of all wasps that came near enough to see the bees. I then offered various dummies instead of dead bees only. My standard set-up was: one freshly killed bee, one dry, dead piece of *Calluna* wood about the size of a bee; one dead bee that I had kept for some time in alcohol and, subsequently, ether, and then dried so as to deodorize it; one stick that had been shaken with live bees, which had given it bee scent; and finally a deodorized bee that was subsequently given scent by shaking it with live bees. The live bees may not have liked being shaken, but as far as I could see they did not come to serious harm and I had no compunction in forcing them to lend me a hand.

In the course of that season I got some quite satisfactory results. There was very little variation in the responses of the hundred or so wasps that visited my exhibit. All dummies were approached and 'hovered at' in equal numbers. All those carrying scent were 'captured'. But the scented sticks were never stung, while the re-odorized bees were.

These results clarified the story. The first response was

indeed visual. When hovering, the wasp checked the scent. After the prey was captured, new stimuli determined whether it would be stung or not and, for this, either detailed visual stimuli or touch stimuli were necessary.

The nature of the first visual stimulus situation was relatively simple: any moving object roughly the size of a Honey Bee was given closer inspection. The size could vary within wide limits: even quite large Bumblebees were inspected, so were small flies half as long as a Honey Bee.

Remembering that the wasps were able to smell a single bee at a distance of several yards, I was amazed to see that no hunting wasp ever approached my scented dummies from more than a foot away. I even put ten bees in an open tube and concealed this tube in the Heather; although their scent was so strong that even I could easily smell them a foot away no wasp was ever seen to respond to them! The difference with the behaviour of the retrieving wasps was striking indeed. Obviously all these hunting wasps were 'set' for was a visual stimulus, and only after receiving this were they 'set' to respond to scent. This impressed me strongly; the results were so clear-cut. Yet I found it all incomprehensible. Later I was to discover that this is quite a common character of instinctive behaviour chains—which, of course, does not make it easier to understand, but which shows that we have to do with a general phenomenon.

This same tube test could be repeated with live bees, so that, apart from their scent, their buzzing could be perceived at some distance. I did this a couple of times, but again no wasp ever reacted to this, and in fact I don't think they can hear at all.

One final point I settled before the season was over: when the wasp had examined the prey's scent and decided to jump at it, what stimuli guided this jump? The flashlike movement, its unfailing accuracy looked very different from the hesitating, clumsy approach by a retrieving wasp and it was hardly imaginable that scent guided the jump—almost certainly the wasp relied on eyesight. I now prepared for the wasps a slightly more complicated arrangement. A scented stick was offered about an inch upwind from a non-scented one. Wasps approached them, hung in the air as usual, and then jumped. But they never seized the scented stick; they always took the other one, which was nearest to them. Of course this meant that, while the jump was triggered off by the scent, it was guided by the sight of the prey.

The outcome of these tests gave us plenty to think about. For instance, the fact that the wasps always hovered at a distance of 3–4 in. from the prey, with very little variation, showed that they must have some rather accurate way of judging distance. But then they must be able to see a difference, at least in size, between a Honey Bee and, say, a large Bumblebee. Yet they never learned to distinguish an abnormal prey or dummy by sight, for they kept reacting to my models, even to odd large ones which I occasionally used. This was in striking contrast to the truly amazing capacities to learn landmarks that we found in the homing experiments. We had to assume that at certain times the wasps were 'set' to learn certain things, at other times they were not.

Thus we were again confronted with a certain rigidity of behaviour, with certain restrictions in what the wasps could do or learn to do at a given moment. Yet in other respects we found almost incredible flexibility. For instance, the movements which the wasp made when turning a bee until it was in the correct position to be stung were highly variable; of course, they had to be, for they had to counteract the bee's violent and unpredictable movements.

It is, of course, the very fact of the lack of flexibility

of so many aspects of their behaviour which kept up our confidence in the ultimate effectiveness of this type of research. The ground for this confidence seems to be the fact that the limitations of an animal's behaviour become increasingly obvious the more we know. The limitations may not always be obvious, yet they are none the less real. To mention a few instances: a wasp will never rake the sand away by pushing it forward or by using its jaws, although it could easily do this as far as muscular equipment is concerned. But when it has to transport a small pebble, it will always try to hold it with the mandibles and never, for instance, push it away with its body.

It had taken us five summers to build up this picture of the life of the bee-killers—admittedly a long time. But this type of work always proceeds slowly, with setbacks by bad weather, the lack of control over animals living in the wild, and so many other handicaps inherent to field work. Yet it would have been absolutely impossible to do these things in the laboratory; it was a matter of doing it in the field or not doing it at all. And there were compensations. One of them was the opportunity the field-worker often has of seeing the entire behaviour pattern of a species. Thus Van Beusekom could extend his studies to an investigation of the wasps' feeding behaviour—their responses to flowers, their scent and their colours. This alone makes a fascinating story, but it will not be discussed here.

Towards the end of the 1930s the wasps began to decline in numbers. It had gradually become clear that *Philanthus*, normally a rare species in Western Europe, had increased enormously in numbers in the late 'twenties. Our attention had been called to them during the peak years round 1930 and it was their abundance that had tempted us into this experimental study. When numbers began to decline after 1935 we soon reached a stage where large-scale experiments were no longer possible. Consequently we turned our attention to other animals. We had by that time discovered so many possibilities for further research that I was determined to hold on to Hulshorst as a training ground for my students. We never had any reason to regret this decision.

CHAPTER FIVE

SWIFTER THAN SWIFTS

While spending the hottest parts of our summer days watching *Philanthus* and scanning the sky for our wasps to return to their homes, we could not help noticing our neighbours, the Hobbies. Like the wasps, they often appeared as tiny black dots against the dazzling blue sky. High above the ground, scarcely visible with the naked eye, they were leisurely hunting insects. Hobbies are undisputed masters at this. Apparently without any effort they flew above the plains, occasionally made a few quick wing strokes, and then a fast turn with one talon suddenly shooting out to seize some small object. Then they resumed sailing and, bringing the foot to their bill like a parrot, nibbled at whatever they had got. This lasted one or two seconds at the most; then they dropped the remains of the prey and began to look out for a new quarry.

Following them through our field glasses (and thereby missing the return of some of our wasps) we sometimes saw them drop small dark objects and, following these until they fell on the ground we succeeded several times in retrieving them. In most cases they were parts of Dung Beetles—which rather surprised us at first since we did not then know that Dung Beetles spend many hours of warm days flying about high in the air. (I still do not understand what they are doing up there.) The remnants we found often consisted of the thorax with head, legs and elytra, and even wings, still attached—in fact, the whole beetle without its abdomen, which obviously was all the falcons were interested in. The hapless, mutilated creatures were still alive and crawled round in a clumsy way. If it were not for this cruel aspect of the whole business, the elegant little falcons feasting on a superabundance of easily acquired food would have made a very pleasing scene of peace, plenty and prosperity.

Sometimes the Hobbies were after more difficult prey. Instead of the short effortless chase after the sluggish beetles, we often saw prolonged, determined swoops over distances of 100 to 200 yards, ending in a quick, zigzagging dash. In such cases we could often see that they were catching Dragonflies. After seizing them they usually broke off the wings and ate the rest.

But even this was not the Hobbies' record achievement. Real demonstrations of their tremendous powers of flight could be seen when they were after their main type of prey—small birds. Hobbies catch large numbers of Skylarks—a difficult enough species to take—but they can do still better; for Swallows and even Swifts are by no means safe. To see a Hobby whizz straight down from perhaps 1,000 ft., wings almost folded alongside the streamlined body, and grab a Swallow in passing so that you can hear the impact 100 yards away is really an experience. The rush of air as the Hobby shoots through it like a meteor is almost frightening. Swallows and other birds often show their deadly fear of their terrible enemy by dashing for cover even when the Hobbies high up in the air start a harmless game among themselves by diving at each other.

Once I saw a Hobby attack—and miss—a flock of Crested Tits that were feeding in some young Birches at the edge of the wood. All I saw was the last stage of its dive, and as it passed me like lightning, missing me by a mere 6 ft. so that I felt the wind, I had a split-second

Fig. 8.   Hobby chasing a Skylark.

view of the slate-grey back, the black dots on the white underside, the red 'trousers' and the bright yellow talons. The tits had all frozen still in the thick of the Birches and when I walked up to them they almost allowed me to touch them—too scared of their winged enemy to venture out into the open.

Because of its tremendous hunting feats, and because of its beauty and rarity as well, the Hobby had a kind of romantic appeal to us. They are late nesters. While all the other birds of prey of the region had fledged when we put up our camp in the first days of July, their eggs had only just hatched at that time. Further, since they were not as dependent on fine weather as our digger wasps, they made a good alternative target for field observations, to which we could turn on cool days or in the early mornings and in the evening. Because of this they were very useful for boosting morale.

In the course of the years we found many Hobbies' nests. Usually there were two, about a mile apart; in some years we knew three nests. These, probably old Crows' nests as a rule, were always in the crowns of Pine trees. They were rarely in dense woods, but usually in small plots of tall timber bordering the open plains. Once we built a tree hide 3 ft. from a nest, but although this did give us some marvellous close-up views of the birds, it did not provide a very good opportunity for watching their behaviour; because we saw them only when they were actually on the nest. With raptors, however, much of great interest happens away from the nest and we therefore changed our tactics and built ground hides of branches and moss some 50 to 100 yards from the nest. From there we often had a wide view and with our field glasses could still see very well what was happening on the nest itself.

Our observations were usually made in spells of 2–6 hours. Sometimes we observed through a whole day from dark to dark. Our most systematic observations, of which we kept detailed notes, covered in all over 500 hours and were done at five nests.

These observations were rather strenuous, for hours could pass without much of interest happening. Particularly when the eggs had not yet hatched, the female would just sit for hours on end without stirring. As in other raptors, the female alone broods. The male appeared only a few times a day, but when he appeared we saw a dramatic event—the 'pass' of food. He broke our monotonous watch by calling loudly in the distance, a clear 'quew—quew—quew—quew!' The sound of it would rouse the female as well as us. She jumped from the nest and flew out to him with strong, quick wing beats. She often met him a couple of hundreds of yards from the nest and, while he slowed down, she threw herself right on her back, in mid-air, and with her talons took the prey from his. She then flew back to one of her 'plucking stations' not far from the nest (each female used a few strong, accessible branches as such) and plucked and ate her meal. The male circled round a few times, or he might even settle in a nearby treetop for a while, but he soon flew off again.

All through the summer we saw relatively little of the males, which was a pity, for they are smaller and of a more slender build, more 'dashing' than the females; the most beautiful creatures one can imagine. Even when we climbed the nest tree the male would keep away, at the most calling loudly high above the wood, while the female often made furious swoops at us. After the meal the female went back to the nest and we knew that we were in for another stretch of one or more uneventful hours.

This division of labour is an interesting feature of many birds of prey. With most species, such as the Sparrow Hawk, the Goshawk and the Kestrel, it lasts until the young are about half-grown. Then the female need no longer guard nor brood them, and she starts hunting on her own, often bringing in slightly larger prey than the male.

All through the season we could observe this beautiful 'pass'—an aerial manoeuvre we never grew tired of watching. Some observers report that they have seen the male Hobby drop the food and the female catch it in mid-air, but we never saw this, although we observed more than a hundred passes.

Once I saw a highly unusual and very amusing pass. On my way to a small patch of old Pines where I had a hide commanding a view of both a Hobby's and a Kestrel's nest situated about 100 yards apart, I saw the female Hobby fly off in a straight line indicating that she had spotted her mate approaching with food. Expecting to see a pass, I followed her through my glasses. After having flown about 500 yards she threw herself on to her back and made contact with the male—or so I thought. But at the same moment I saw that the other falcon was the male Kestrel. A Kestrel is not used to this procedure —it hands its prey over after alighting on a branch—and quite understandably this male refused to let go. But the Hobby held on and the poor Kestrel was dragged down, screaming at the top of his voice. Down and down they went. They disappeared behind a sand dune and when they emerged again, a second or so later, the Hobby had the mouse and made straight for her nest, while the Kestrel followed lamely, to meet his wife empty-handed. She followed him, begging loudly, for several minutes— rather tactlessly, I thought.

With the hatching of the young, early in July, watching became increasingly interesting. At first the mother still covered her young. When the male came, she took his prey as before and flew to one of her usual branches. While plucking what feathers the male had left she often took mouthfuls of meat herself, but before long she flew off and took the prey to the nest. And there the fierce predator changed into a gentle, loving mother. Carefully she tore tiny particles of meat off the prey, bent down and patiently held it in front of her young. Through our glasses we saw them for the first time: tiny, wobbly heads, covered with pinkish white down made faint and ill-directed pecking movements in the direction of the mother's bill. After a few, or often many, failures one got hold of the food and greedily swallowed it, falling down into the nest cup during this strenuous effort. Again and again the mother offered food and we watched all the young getting several turns, until, after about 15 minutes, they ceased to respond and dozed off. The mother then swallowed the rest, picked up morsels that had been spilled and settled on the young, to take a nap herself.

The intervals between feeds varied considerably. Most of our pairs fed exclusively on larger prey such as song-birds. In 41 cases we could time the exact interval between two consecutive feeds. The average was 77 minutes, but the variation was enormous—from 4 minutes to 185 minutes. Even longer pauses did occur; once we started observing at 1 p.m. and the first feed was seen at 4.53, almost four hours later! Some of our pairs fed insects such as Dragonflies and then the rhythm was quite different—a prey every few minutes.

When the young grew up, numerous changes occurred in their behaviour, in that of their parents, and in the parent-young relationships. We made a point of observing these as accurately as possible and made many interesting discoveries, some of which I will describe.

When the young were about ten days old, they began to turn their attention to the prey instead of waiting for the mother to present a neat little morsel. They now began to tear at the prey and soon they were quite a menace to their mother and often knocked her over. She was at first reluctant to give in, but she was forced to do so in the end and the young got their way; soon she just dropped the prey on the nest and hurried off.

At about this time the young also began to show their true nature as raptors, by becoming interested in small moving objects. In spite of the mother's efforts at nest sanitation, flies soon buzzed round the nest. They were an ever fascinating source of wonder to the young, who followed them with their eyes, under the most curious contortions of their necks and bodies. Tits and other small birds feeding in the nest tree attracted concentrated attention. But their idea of 'prey' was still very vague: similar and quite intensive reactions to the downy feathers floating away when the young preened themselves proved that it was the mere movement that fascinated them—there was as yet no question of 'prey-recognition'. Nor were there any attempts at catching prey or even movements 'in vacuo' as have been observed

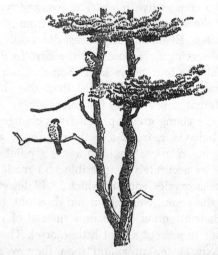

Fig. 9. Fledgling Hobbies on the look-out.

in other raptors. By this I mean hunting behaviour carried out in the absence of prey.

In birds of prey amusing examples of this have been observed. Young Honey Buzzards—eagle-like birds the size of Buzzards, which prey on wasps and their broods, digging up nests with their feet—do a large amount of premature digging while still in the nest. Young Secretary Birds perform the typical 'dancing' movements of killing a snake many days before they leave the nest, a wonderful performance, which has been filmed in detail by Dr G. J. Broekhuysen of Cape Town. That Hobbies do not indulge in similar premature hunting 'games' is probably because they hunt in flight. Once they fly, young Hobbies do perform hunting games, as do Kestrels. I will describe them later on.

A most interesting time began when the young were about 4½ weeks old. They then began to leave the nest and took up positions a foot or so outside it. Although they were by then able to fly, they did not do so unless disturbed. Every day they extended the range of their 'walks' in the treetop. At this stage it was amusing to watch how imperfect was the coordination of their movements with distance perception. We once saw a young Hobby step on a branch a few inches above its perch. It looked at it, lifted a foot, but in putting it down missed the branch and all but tumbled over. It tried again and stepped short again. This repeated itself fourteen times! At the fifteenth attempt it had corrected itself and, fluttering clumsily, it stepped, or rather crawled, on to its new perch.

Soon after this the young birds began to fly, over short distances at first. They took up roosts to which they resorted for periods up to a day. The parents (for both were now hunting for the family) now took their prey to the young, usually alighting near the first one they encountered. This made the young shift their position further and further in the direction from which the parents usually approached them, which, with our Hobbies, was always the north-west, where occurred the fertile pastures and villages along the coast of the Zuiderzee which were their favourite hunting grounds. Often we could observe striking cases of the parents learning the favourite perches of the young. When they had fed a young bird a couple of times on a particular perch, they would return there next time even though the perch might be deserted and the young were on other perches in plain view, screaming for all they were worth.

The young by this time had attained an amazing delicacy of perception. They showed by their behaviour that they distinguished Hobbies from Wood Pigeons and Sparrow Hawks, for they never reacted to these species by begging as they invariably did to their parents. But in the beginning they did mistake a Kestrel for a Hobby; the similarity between those two species, of course, being great. Later, we saw to our amazement that they had learned to recognize each other and their parents individually, for when a strange Hobby, young or old, approached their trees, they sleeked their feathers and never uttered their begging calls. Yet at the same time they were unable to understand that a parent without prey was not worth begging at—they started their begging cry every time one of the parents showed up and even followed them persistently in the air, trying to take over a prey that they could see was not there!

Aerial 'passes' between parents and young began to occur a few days after the young had started to fly. Here again the clumsiness of the young birds' movements was striking; they often had to try several times before they got hold of the prey, forcing the parents to turn and meet them again or hover awkwardly with the prey dangling under them, waiting for one of them to manoeuvre into position again after a fruitless attempt. The struggles of the young were downright pathetic when the parent had an insect; time and again they would try to grab the tiny prey, sometimes making as many as seven attempts.

The parents also revealed wonderful powers of perception. Like most other birds they had a keen ear and amazing powers of sight. We often saw the female dive off the nest as soon as she heard her mate calling 1,000 yards away, when it was quite impossible, on account of the forest, to see him. But purely visual responses were even more common. They distinguished Kestrels from Hobbies at great distances; the incident described above was the only exception we saw. Sparrow Hawks they ignored, but the extremely similar (though larger) Goshawk they attacked furiously. This, incidentally, cost one of our Hobbies its life: when it swooped down on the Goshawk the latter just turned round in mid-air and killed it outright—an incident observed by my friend

George Schuyl, who joined us in watching the Hobbies for several summers.

The adults could also distinguish strange Hobbies and were better at this than the young. Any stranger was recognized a long way off. They flew towards it in a strikingly energetic and determined way and we saw these aggressive flights start when the trespasser was over 500 yards away! They were not dependent on the other member of the pair being present so that the attacker could therefore know that a distant Hobby must be a stranger.

One year we spent quite a time observing Hobbies that were hunting Dragonflies. On numerous occasions we could measure the distance from which they saw them, again because their flight showed their intention unmistakably and ended in the capture of a Dragonfly. We measured distances up to 200 metres, which is more than twice the distance in which we ourselves could see them with the unaided eye. If we assumed that the maximum thickness of the insect's body was 2 cm. (an overestimate if anything), then the visual acuity, as expressed by the smallest angle perceivable, is about 21″, which is actually about twice as small as that of an average human being expressed in the same way.

Another measurement we tried to do was that of the speed of flight. Through our field glasses we could follow a Hobby until it was about 3 kilometres away. By comparison with landmarks it passed it was possible, under ideal conditions, to check distances up to this maximum. At about that distance, 2.5–3 kilometres away, there was an area where the Hobbies captured many Dragonflies, which they then carried back to the young, not far from where we were watching. We could just see them swooping down on their prey and immediately afterwards began to time their return flight, which was a relatively flat glide. Our two most reliable measurements were 60 and 69 seconds. This gave us a speed, over 2.5 kms., of at least 150 km. an hour. We were convinced that the Hobbies' maximum speed is very much higher than this, for the glide, fast as it was, was much slower than the Hobbies' swoop, straight down, on a Swallow.

A few times we visited the Hobbies' area during the night. Even on moonlit nights there was complete silence. On several occasions I started my watch one to two hours before sunrise. I vividly remember one clear morning in July, when I went into the hide 90 minutes before sunrise. When I walked under the nest there were no signs of alarm. Nightjars were singing until 4.30, but already 15 minutes earlier I heard the first calls of the Hobbies. At 4.46 a Chaffinch started singing, four minutes later the first Wood Pigeon began to coo. At 5.53, about half an hour after sunrise, the male Hobby brought its first prey, at 7.20 a second. Since the Hobbies were awake at least an hour before sunrise, it is possible that they do some hunting in the morning twilight and this may explain some of our facts about their food which I will mention later.

After what I have written I need not stress that the Hobby is a real master of flight, capable of high speed and of catching some of the most able fliers in the Animal Kingdom, such as Dragonflies and Swallows. When we saw how clumsy was the flight of the young Hobbies in the beginning, we were naturally anxious to see how long it would take them to acquire sufficient dexterity to earn their own living. Luckily the open country and the conservatism of the young, which kept them near their birthplace until at least September 10th (5½ weeks after fledging) gave us the chance to follow them through this long and most interesting period.

Soon after fledging, the young began to make spontaneous aerial trips, flying in their unsteady manner for 50 yards or so and then returning. Their first clumsy attempts at landing were amusing to watch; like other birds, they had more trouble alighting than in taking off. However, they improved rapidly, perhaps through practice, and as they grew up they ranged further and further. Towards the end of August they could fly 1,000 yards and more and quite often made trips of 500 yards. Also they spent more and more of their time in the air.

A week or rather less after fledging we usually saw them catch their first insects in the air; or, rather, we saw them try to, their first attempts often ending in failure. Sometimes two went after the same beetle; that ended in a collision, after which the two youngsters would tumble down in a flurry of wings, dropping many yards before they resumed control, while the unconcerned beetle buzzed stolidly on his way—until caught by the next Hobby.

Dragonflies were too difficult for them at first, but once they mastered the beetle-catching technique they began to go after this more difficult quarry as well. Before the end of August Dragonflies were no longer safe from them either. But never did we see them make any attempt to chase a bird. The last day we ever saw one of our Hobby families was on September 9th. They must have left the area soon afterwards. I wonder whether young Hobbies are independent when they start on their autumn migration.

The Hobby technique of catching birds looks extremely difficult and it is probably no accident that young Hobbies have to be fed by their parents until their powers of flight have reached a high perfection. It seems likely that their continual practice does develop their skill. Not only are they forced to do their utmost to master the trick of the 'pass'; they also play a kind of hunting game with each other. Now and then, while cruising or hawking insects high up in the air, they would suddenly begin to swoop down on each other, just as if they were hunting birds. They did not actually try to seize or even touch their brother or sister, but they sometimes stretched out a yellow talon in passing. We never grew tired of watching them at this game, and by observing their improvement day by day we began to realize how wonderfully expert the adult Hobby's swoop at a Swallow really is—quick, smooth and highly accomplished. Like everything that is really well done, it gave the impression of being quite effortless.

Several times we could, by a lucky coincidence, observe the effect the swooping movements of the Hobbies had on Swallows. Loose flocks of approximately 50 Swallows and Martins often passed the sands on their way south, feeding leisurely at an altitude of about 300 ft. They never showed that they were in the least concerned about the Hobbies, hawking about 1,000 feet higher up. In view of the many proofs we have had of the extremely keen eyesight of birds, we have not the slightest doubt that these Swallows saw the Hobbies perfectly well. When the latter started their swooping games, shooting through the air like living arrowheads, the Swallows dashed for cover. Although the Hobbies did not come anywhere near them, staying high above them, the Swallows dropped like stones and disappeared into the Pines. From field observations such as these one can get a fairly good idea of the way birds recognize birds of prey. In this example the Swallows showed by their behaviour that the Hobbies' way of moving was to them an important stimulus. The fact that they did not show any particular behaviour while the Hobbies were quietly cruising does not mean, of course, that they had not recognized them for what they were—we simply cannot say.

I have made observations which show that some birds at least are frightened even of swooping birds that do not resemble birds of prey in the least. Gulls and waders, for instance, can panic when a harmless bird such as a Godwit, in an exuberant mood, drops down like a stone only to settle near them. On other occasions it is obvious that the particular shape of a bird of prey is recognized. The well-known examples of birds showing fear responses to Swifts when they first arrive in spring must be due to the fact that Swifts, like falcons and hawks, have short necks. I have also seen Tits mob a flying Nightjar; and again the Nightjar has this same character in common with birds of prey.

Several people have done experiments with cardboard models of various shapes which they sailed over various birds. In these experiments the birds tested—gallinaceous birds, ducks and geese—showed alarm at all models that had a short neck. Curiously enough, young white leghorn chickens did not show this response. The observations of Smith, Edwards and Hosking on the responses of song-birds to Cuckoos show that a Cuckoo is treated in quite a distinct way again. While the form of a Cuckoo shows superficial resemblance to that of a Kestrel, the latter is certainly not attacked the way a Cuckoo is. And this again is rather puzzling when one thinks of the Tits' attack on the Nightjar, for a Nightjar's silhouette certainly is more like that of a Cuckoo than that of a falcon. Yet I do not think Tits attack Cuckoos.

Anyway, there is still much to do on the subject of how birds recognize birds of prey and why they are so variable in their responses; why, for instance, a Harrier may be furiously attacked on one occasion and completely ignored on another occasion, though he may have been hunting both times. Systematic field observations, followed by experiments, might yield much of great interest.

The Hobbies are not the only birds in which the young remain dependent so long; it is also typical of Terns, and of Owls—both groups which have feeding habits requiring great skill.

In the course of several years we collected some information about the Hobby's menu. Although it does not allow us to make more than a crude estimate of the proportions in which various species are represented in the diet, it was worth collecting what data we could, because in most Hobby habitats their food cannot easily be assessed since most of the remnants such as pluckings and pellets are lost in the thick undergrowth. Our Hobbies all nested in Pine plantations with hardly any undergrowth at all and we found more preys than is usually possible. Sometimes we could determine the prey by actually witnessing the kill. On a few other occasions we could see what prey the mother fed to the young on the nest, although, watching from the distance we did, we could rarely say more than that it was a bird or a small mammal. But through our continuous watching we found that each pair had a few preferred branches where they did most of the plucking of the prey and by inspecting the ground under these plucking stations once or twice a day, and collecting feathers, pellets, etc., we gradually accumulated the following data.

Of 171 large prey, 135 were birds, 27 were mammals, and 9 were either one or the other, but we could not be sure which.

The 135 birds were: 39 Skylarks, 18 Swifts, 12 Starlings, 9 House Sparrows, 9 Swallows, 7 Linnets, 4 Yellow Hammers, 4 House Martins, 3 Great Tits, 3 Yellow Wagtails, 3 Tree Sparrows, 1 Wood Lark, 1 Crossbill, 1 fledged young Cuckoo, 1 Whitethroat, 1 Pied Flycatcher, and 19 unidentified birds. The 27 mammals were: 6 Moles, 3 Shrews, 2 Voles, 2 Wood Mice, and 14 unrecognizable.

The most important prey, therefore, is the Skylark.

The German name of the Hobby (Lerchenfalke=Lark-falcon) is very appropriate. Our data gave us a rough idea of the total consumption of Skylarks by one pair of Hobbies. They spent about 5 months with us, say 150 days. We know that on the average a female with 3 young receives a prey about 11 times a day, which means a ration for each of 2¾ animals the size of a song-bird. The family uses, again roughly, 2 rations a day for 90 days, and 5 rations a day for 60 days, which makes $480 \times 2.75 = 1,320$ prey animals. If one quarter of these are Skylarks, this means that one family of Hobbies kills 340 Skylarks per season. This figure gives scarcely more than an impression of the scale of the toll levied by such a predator. Far more accurate figures are known of the Sparrow Hawk and the small song-birds it preys on, and of the Tawny Owl and the small mammals which are its food. Thorough investigation, not only of the food, but also of the range of the hunting area and of the population of the prey species have led to a fairly accurate estimate of the part these predators take in the total mortality of the prey species. Our scanty data on Hobbies did not carry us anywhere near that goal.

Another striking feature of the food list is the high proportion of Swifts. We found them as pluckings always during periods of prolonged cyclonic weather, when the Swifts were probably in poor condition. Some, but by no means all the Swifts were fledglings. To our great regret we never saw a Hobby in the act of catching one, although we saw them chasing Swifts as well as Swallows.

The high number of mammals also surprised us, for the hunting method does not seem to be well adapted to picking up mammals from the ground. The incident of the Hobby robbing a Kestrel might suggest that this is the way they acquired most of these mammals, but we were sure that this is not so, for we actually saw

Hobbies snatching mice from the ground on more than one occasion. They did it much in the way a Kestrel does it, by gently gliding down from above and unhurriedly picking it up. Peregrines, on the whole also aerial hunters like the Hobbies, occasionally take mammals in the same way.

What little we discovered about the Hobbies' insect food was equally interesting. As I said before, they take huge numbers of Dung Beetles and Dragonflies. But under their roosts we found numerous pellets, light brownish-buff in colour, and containing very few remnants of either. Inspection of the fragments under the microscope showed that they consisted of scales and other parts of moths and, much to our surprise, they were typical of the scales of Pine Hawk Moths! These were abundant in the pine woods, but they never fly by day. We must assume that the falcons got them at dawn and at dusk mainly, when the moths fly in great numbers. Twice we found the torn-off wings of Pine Hawk Moths under the plucking branches.

We found these pellets containing remnants of insects year after year, sometimes dozens of them in a month. At first we were rather puzzled by them, for, while they showed that our Hobbies ate insects in considerable numbers, we never saw them feed any insects to their young, not even on warm sunny days when insects are active. Yet this had been reported of other pairs. Later we found some pairs that did bring insects to the nest. They did this only on sunny days; on cold days probably too few were available. Yet it seemed as if, when the insect supply dropped below a certain level, the falcons stopped taking them altogether—they either brought in many insects or none at all.

We were also surprised to see that the different pairs behaved so differently: on the very day when one pair would bring in nothing but birds another pair would concentrate entirely on insects and feed, for instance, 70 Dragonflies in four hours. Both pairs were equally far from Skylark-country and in both territories Dragonflies were abundant. We could only conclude that the pairs themselves were different in their tastes.

Another striking point was that the male never gave an insect to the female first, as it did almost invariably when carrying larger prey. This was because the female was not interested when the male had a Dragonfly. We have definite proof that she could see the small prey perfectly well and also that neither the behaviour of the male, nor the weather itself, determined her behaviour. Here we got a hint of the complexity of the factors determining such things.

Of course, the value of these observations should not be overrated. On the whole, they led to more detailed descriptions rather than to conclusions about the causes of behaviour and I can well imagine my readers wondering why we probed into these seemingly trivial matters at all. Such observations act as a kind of survey of the kind of things animals do and which one would like to explain. They provided excellent practice for my students, and they kept up morale on cold and wet days. And we did see much of great interest: the many traits of behaviour peculiar to a highly specialized animal, wonderful examples of the highest achievement in flight and of visual perception, alongside amazing demonstrations of 'stupid' limitations. We watched the growth of behaviour patterns, the development of parent-young relationships, the influence of the weather on behaviour, and so on. Observing these things in any animal is fun and the species, or the importance of the particular problem studied, are entirely irrelevant when one is trying to get experience in a method of study.

Thus we gradually began to see a number of problems

worth following up, and although the Hobby itself did
not offer very good opportunities for more penetrating
analyses, we will never regret having spent so many
hours making the acquaintance of these superb birds,
swifter than Swifts.

CHAPTER SIX

THE SAND WASPS

The glacial sands of the central Netherlands where we watched the bee-killers were quite different from what they used to be. Before Man began to utilize these poor soils by laying out the extensive Pine plantations one finds there today, heaths covered areas of well over 1,000 square miles. Here and there in these very slightly undulating plains Man had settled in small villages of low, thatched cottages, built in groups round a village green and a little church, and tucked away in the shelter of groups of trees. Surrounding such villages there were belts of arable fields where potatoes, rye, oats and buckwheat were grown, and which were protected from the inroads of Roe and Red Deer and Boar by dense hedgerows. Somewhere in the arable belt a windmill stood on a well-exposed site.

Sheep were kept in considerable numbers and each morning the flocks, guarded by a shepherd, went out to graze on the heaths. Except for an occasional pool in a shallow depression, where green vegetation and brightly coloured flowers broke the monotony of the heath, the carpet of *Calluna* was a dull brownish colour for most of the year. Only in August this turned to a lovely purple, and then millions of insects of many different kinds busily fed on the delicious nectar.

With the planting of Scots Pine this country changed.

The Pine plantations soon grew up to large complexes of rather dull, tall timber. Millions of seeds were thrown out by these trees over the surrounding country, so that in suitable areas scattered groups of naturally grown Pine trees began to spring up, beautiful, twisted trees with wide crowns, quite useless but very attractive.

To prevent the spread of fires, which now became much more dangerous because of the plantations, strips of soil were cleared of vegetation. The bare soil so exposed offered suitable homes for numerous burrowing insects. Prominent among them was a digger wasp called *Ammophila*, one of the caterpillar-killers that have been famous ever since Fabre described the way in which they paralyse their prey by stinging the nerve centres in the larva's body.

These wasps look very different from *Philanthus*. They are long and slender, black, with a red area at the base of the abdomen. The females dig shallow burrows consisting of a vertical, narrow passage which opens into a chamber 1½–2 in. below the surface. Here they store caterpillars, which provide the food for their larvae. On one of these caterpillars, the first to be carried in, the wasp lays its egg. The wasp's larva devours the caterpillars, then pupates, and lies dormant in the burrow all the winter. Next summer the wasp emerges and during June, July and August many of them roam over the heath—the males, carefree, living on the nectar; the females digging, hunting and laying their eggs.

Ever since we started our studies on the homing of *Philanthus* we were aware of the need to examine the orientation of *Ammophila* as well; for the use of landmarks, so easy for a wasp that lives in a bare, open habitat and that carries its prey home in flight, might not be so easy for a creature like *Ammophila* that lives among knee-deep Heather with few outstanding landmarks and that has to carry its heavy prey home walking over the

ground below the Heather shrubs. It seemed quite possible that *Ammophila* had to use an entirely different
method of homing.

An accident made us tackle *Ammophila* before we had
finished with *Philanthus*. In our summer camps we had
developed what we called the 'slave system'. Young undergraduates volunteered to join us in the long vacation
and to help us with the field work of the graduates, thus
gaining experience as well as doing a useful job. One
summer Van Beusekom, then engaged in the *Philanthus*-
work described in Chapter Four, fell ill and had to forgo
a whole season. Two slaves had already arrived and were
keen to start work. Discussing emergency plans, they expressed an eager desire to tackle *Ammophila*. And that
was how G. P. Baerends and Miss J. van Roon (now
Mrs Baerends) began a study, the results of which grew,
over a period of seven summers, into a most wonderful,
exciting and in some respects unique story. My pride in
this matter is mixed with a great deal of embarrassment,
for in spite of my own preoccupation with homing, the
Baerendses went their own way and studied several
other aspects of the Sand Wasps' life, and discovered
many very remarkable things.

They began by spending a number of sunny days on
a bare path running across the heath, selecting a stretch
of a dozen yards or so where many wasps were found
at work. The species they found here was the smaller
of the two most common species, *Ammophila campestris*
Fur., as it was then called. This Sand Wasp, although
much more difficult to observe than the larger *Ammophila sabulosa*, was chosen because it nests in groups.

Before the Baerendses could recognize individual
wasps, they could not see more than disconnected fragments of behaviour. These, however, were interesting
enough in themselves. Some wasps could always be observed digging at their nests. This they do with two

'tools'. With their strong, pointed mandibles they bite into the hard, sandy soil. When the sand has been worked loose, they rake it backwards with shovelling movements of their front legs. These carry a kind of broom made of stiff bristles. Unlike *Philanthus*, these wasps do not leave the excavated sand about, but they take it 'under their chins', by bending the head against their thorax, and then carry it away in flight. They drop it a couple of inches away. Thus the sand never piles up round the nest entrance and because of this the nest is extremely difficult to find.

In good weather the digging of a burrow takes about 45 minutes. As the nest becomes deeper, the wasp's body disappears further and further into it, until it stands on its head, only the tip of the abdomen showing. When the shaft has got to about this stage, the wasp digs a chamber. This provides just enough room for the larva and some caterpillars. It is just large enough for the wasp to turn round, but she usually comes out tail first every time she goes in head first, and vice versa.

When the nest is finished the wasp 'closes the door'. That is, she collects little pebbles or bits of wood in the neighbourhood and pushes them into the shaft. As long as the nest is empty, it is closed only provisionally when the wasp leaves; she uses bits of wood, pebbles or little clumps of earth, and this nest can be recognized as a shallow hollow in the surface. Once there is something in the nest—an egg or a larva—the female works sand among the pebbles as well and rakes sand on top so that the entrance is completely hidden from view.

In the course of a day, wasps often arrived with caterpillars. These might be as big as the wasps themselves and preys of that size have to be dragged home. With lighter loads the wasps flew over short distances, but rarely more than a couple of yards at a time.

A wasp carrying a prey seemed to know exactly

where to go. Upon emerging from the Heather she often walked in a straight course along the path, stopped and dropped the caterpillar, and apparently without any hesitation began to shovel the sand away. In nine out of ten cases the wasp had obviously started at exactly the right spot. Soon her raking uncovered the pebbles, which were then taken out one by one, and within a very short time the door was opened. We will later see how the wasps managed to find their nests so accurately.

After opening the nest, the wasp dived into it. Often she first cleared out the sand that had dropped down when she removed the pebbles. Then she turned round, until the tip of her abdomen was just above the nest entrance and, seizing the prey in her mandibles, dragged it backwards into the chamber. After a while she came out again and closed the entrance again before starting out once more.

It did not take the Baerendses very long to see these and many other details of the wasps' behaviour. But from these odd bits they could not put the whole sequence of events together as they followed each other in the life of individual wasps. It was practically impossible to follow individuals for more than an hour at best, for when they left the nest site they immediately disappeared into the dense shrubbery where one lost sight of them almost at once. Therefore, the next thing the Baerendses did was to give some wasps colour-marks by which they could recognize them whenever they saw them. Tiny dots of a quick-drying paint did the job. Further, they marked each nest as soon as they found it by sticking a piece of thin wire into the sand an inch or so from the nest entrance. The wire was bent into the shape of an upturned 'J', of which the free end was an inch and a half above the entrance.

Then they began keeping a record of what exactly each marked wasp did—a tedious job demanding continuous vigilance and great perseverance, for these wasps are far less conspicuous than the bright yellow *Philanthus*. But their painstaking work was worth the trouble, for soon an almost incredible story emerged. They found that each wasp began, after digging a nest, by capturing one caterpillar. This they stored in the nest and, staying down for 20–100 seconds, laid an egg on it. They then left this nest for the time being and either just went loafing in the Heather or, more often, started digging a second nest. Or, equally often, they went to a different site on the path and opened up an already existing nest. Sometimes they just slipped in for a moment, shut it again and left altogether; but at other times they left for a short time only, returning to this nest with a prey. This they then stored, but now they stayed down for, at the most, 10 seconds only. Later it became clear that such a short visit was nothing more than provisioning a larva that had already consumed its original food store.

When the wasp had brought either one, a few or sometimes many caterpillars to this nest, she might return to the original nest again. Here she first paid a visit without bringing in any prey, but after that she usually returned once or twice with a caterpillar. Then she left this nest alone for a second time and often did not return to it for one, two or even three days. In this time she might return to the other nest, or start on a third one.

After some days she again visited the first nest, usually carrying in a number of caterpillars in quick succession. After this she closed the nest in an exceptionally thorough way. Using her head as a hammer, she pressed it forcibly against the pebbles which she had dropped into the shaft—buzzing fiercely all the time. This was the last thing she did for the larva in this nest; it amounted to bolting the door. After this, she continued work at the second or third nest.

Simply by marking nests and individual wasps, and by careful, persistent observation, the Baerendses had revealed an unheard-of-thing. Such a tiny wasp, a mere insect, actually cared for two, sometimes three nests at the same time. She knew the position of each of these among the multitude of similar burrows on the path and, more astonishing even, she knew exactly what to do at each nest! For the story of each nest consisted really of three chapters. The first phase consisted of digging the nest, provisionally closing the entrance, hunting for a caterpillar, carrying it home, storing it, laying an egg on it, and locking the door. Then there was a pause of one or a few days, in which she busied herself with another nest. One day she returned, visited nest No. 1 without carrying in a caterpillar, locked the nest again, returned once or twice with a caterpillar, and after locking the nest once more, left it for another day, again in the meantime working at another nest. Then she came back for a third time, again starting with an empty-handed visit, after which she locked the door, and this time she would soon return and bring in a collection of 3, 4 or even more caterpillars in quick succession. Then she bolted the door and left the nest for ever.

Here, therefore, we have a solitary wasp which, unlike other species, returns to her offspring twice and each time brings in a new supply of food. In the two intervals, between phases 1 and 2, and between phases 2 and 3, she works very efficiently at other nests, which may be at an earlier or later stage. The way in which the various broods were telescoped into each other is variable, as is the time elapsing between the phases at each nest. This depends largely on the temperature. When it is cold, the larva grows and eats more slowly and the mother works more slowly, than when it is warm.

Of course, it took the Baerendses a long time before they knew that this was the general schedule. They had to follow many wasps through many days. This took several seasons' work, because a number of series of observations were interrupted by long periods of cold and rainy weather, during which many wasps and larvae succumbed. As we had already noticed while observing *Philanthus*, the Atlantic climate is not very well suited to these wasps—nor to their biographers.

Usually, weather permitting, a number of wasps were followed through a week or ten days, their activities recorded and, at the end, their nests opened to see what was inside. Later the Baerendses constructed artificial nests of plaster, which allowed them to discover still more of the details. The plaster nest method was simple but ingenious. A solid cylinder of 2 in. diameter and 3 in. high was cut across into an upper and a lower cylinder. In the upper half, a shaft was bored, *Ammophila*-fashion, and a chamber made in the lower half. When a nest was located, a drill was used to take out a cylindrical core of earth which contained the entire nest. The contents of the nest was transferred to the plaster nest, which was then put in its place and covered with the actual top layer of the block of soil that had been removed. Great care had to be taken that this top layer remained intact, that it was placed in exactly the right position, and that the actual entrance fitted that of the plaster nest exactly. After this, the nest could be opened whenever necessary by just lifting the upper cylinder.

It was, of course, a great day when the first wasp, faced with this new contraption, accepted it. From now on, the state of the nests' contents (size of larva, state of food supply, etc.) could be checked at will. And as we will see later, the new technique opened the door to some really exciting experiments.

Observing, simultaneously, the mother wasp's behaviour and the state of the brood revealed that the egg was invariably laid on the first caterpillar, that the second phase began when the first caterpillar had been consumed and the larva was in need of more food, and that the third phase usually began when the second lot had been eaten. By the time the larva ran out of food, she was full grown, and could spin a cocoon and pupate.

These observations on individual wasps were charted as in Fig. 10. Forty-three nests in all were followed from

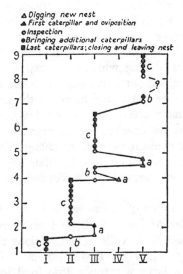

Fig. 10.   Nine days of a Sand Wasp's activities at five nests (I–V).

beginning to end, many of them being successive nests of one and the same wasp. Of several wasps four successive nests could be followed, of others fewer or more; and some wasps could be followed through as many as nine nests. Baerends published seven charts, all differing in detail, but all showing the three phases and the tele-scoping of broods.

As soon as it became clear that these wasps cared for two or three nests simultaneously, each in a different state of development and therefore requiring different amounts of food, one naturally wondered how the wasps knew what to do at each nest. The plaster nest method allowed the Baerendses to find the answer.

As I have shown, the second and third phases often began with a visit on which the wasps arrived empty-handed and left soon after. Sometimes, however, such visits were not followed by a provisioning visit. It seemed plausible to assume that the empty-handed visits (which I will from now on give the name 'calls') served as inspections, during which the wasp received stimuli which determined her behaviour for hours to come, and decided whether she would carry, to the particular nest concerned, many caterpillars, only a few, or none at all.

This could be checked by experiment, because the plaster nests made it possible to change the nest's con-tents at will. The Baerendses, by planning and carrying out these tests accurately, succeeded in solving this question.

First, a number of wasps were watched for several days in order to know their various nests and to be able to predict their 'calls'. Then, various changes were made shortly before a call was expected. For instance, a nest with a larva with only little food left was given a liberal supply of caterpillars. Or a nest that had already received many caterpillars and ought to receive one or two more at the most was robbed of all its caterpillars. Or, again, a caterpillar with an egg was replaced by a larva. In other tests a larva in need of more food was replaced by a cocoon. The tests were usually arranged in pairs, caterpillars, larvae or eggs being exchanged, so that few caterpillars or larvae were lost.

The results were striking. I cannot give them all here,

but will select a few. When the caterpillars were taken out before a 'call', all wasps responded after the call by bringing in more than they would otherwise have done; some wasps brought their total per nest up to 12 or even 13, whereas the normal number is between 5 and 10. When extra caterpillars were given before the third phase, the wasp, instead of storing 3–7 in a row, brought in none at all or, in one case, just one. Thus, changes in the number of caterpillars caused changes in the mother's behaviour. The size of her larva also influenced her; too young a larva never received many caterpillars; also, replacing a young larva by a cocoon made wasps abandon the provisioning visits altogether and, at least if they had already entered the third phase, bolt their nests. Wasps still in the second phase stopped carrying caterpillars to such a nest, but did not bolt it.

These results were so remarkable because they showed not only that the 'call' was really an inspection, and determined what the wasp would do next, but also that the effect of the call could last so long: for instance, when nest 'p' of wasp 'XXX' was robbed of all caterpillars before her call on June 28th, 1940, at 10.30 a.m., she reacted by bringing in one more caterpillar on that day and three more on the next day. The effect, therefore, lasted through the night. In order to grasp the real significance of this fact one has to remember that this wasp was caring for other nests in other stages of development at the same time. The stimuli received during the call did not merely switch her into another general condition, for she did not supply more caterpillars to her other nests. The call made a wasp do a special thing at a special nest. This is a remarkable feat, a true delayed response, and a very much delayed one at that.

Other wasps were subjected to exactly the same tests during provisioning visits instead of during 'calls'. In none of these tests did the change affect the wasp in the least;

she just went on with the job as dictated to her when she had called before.

Imagine what this means in practice. The wasp knows the location of two, three or even more nests, for she may occasionally call at old nests as well. After she has left one nest, she usually makes a call at the nest most in need of supplies at the moment. How she knows this is another puzzle. Then, at this call, she gets her 'instructions' and carries them out. After this she may call at another nest and again acts according to what she finds there. All the time she remembers where all the nests are and, roughly, in what stage they are. The calls are only for checking up.

It will be clear that such a colony of about a hundred caterpillar-killers must make considerable demands on the fauna of the surrounding heath. The observer cannot help wondering how this habitat, which gives the impression of being relatively poor, can supply the scores of caterpillars brought in every day. His wonder turns into admiration when he tries to compete with the hunting wasp and searches for caterpillars himself, for he discovers soon that he cuts a miserable figure.

This made the Baerendses study the hunting habits of their wasps in some more detail. Where did the wasps go? How extensive were their hunting grounds? How did they find their prey? Once they had caught a caterpillar, how did they get home again?

The three species of caterpillar most frequently carried (*Anarta myrtilli, Ematurga atomaria* and *Eupithecia namata*) were all beautifully camouflaged. *Anarta*, a green-and-yellow chequered species, disappears almost completely on the young *Calluna*-shoots on which it lives. *Ematurga* resembles an older, bare *Calluna*-twig, and *Eupithecia* is a pinkish form living on *Calluna* and *Erica* flowers. It was not possible to test experimentally how the wasps found their prey, but incidental observations

showed that a wasp never made for a caterpillar in a
straight line, as one would expect if it had seen it. It
wandered about irregularly until, being one or at the
most two inches from the caterpillar, it began to run in
small loops and circles. This was continued until, as if
by accident, it touched the caterpillar and then the wasp
stung it. It seemed, therefore, as if the wasp discovered
and recognized its prey mainly by scent and perhaps by
touch.

When the Baerendses followed the wasps into the
surrounding heath they were amazed to find that they
did not go at all far from home. When they searched
strips of the heath parallel to the path, they found almost no wasps within five metres from its edge. This
was natural, for this strip was dry and brown and had
almost no flowers. Most wasps were observed in the next
5 metre strip, fewer in the 10–15 metre strip, and so
on, until in the 35–40 metre zone practically no wasps
were ever seen. Thus the hunting area was extremely
small. Yet it was large enough to make it something of
a problem for the wasps to get home. In the struggle
with a caterpillar a wasp often tumbled on to the ground
and found herself surrounded by the *Calluna* bushes
towering above her. Now before starting home the wasp
invariably climbed either a Heather bush or a young Pine
tree. Arrived at the top after a laborious climb, she
turned in various directions, as if having a good look
round. Then she took a long jump, which was always in
the direction of her nest. The weight of the caterpillar
decided how long this 'flight' would be, but it was rarely
more than 2 metres. The wasp then began to walk, stum-
bling and plodding along over the rough ground. She
usually started in the right direction, but often deviated
or even began to make irregular turnings and even loops.
When this happened, she would again climb a 'tree', look
round and make another jump—again in the correct

direction. In this way the wasps got home, often after a long and difficult journey.

After following a number of such wasps on their way home, the Baerendses were convinced that the wasps knew the way, at least when they could have an unobstructed view of the land. They were not just searching at random. The Baerendses now proceeded to experiment.

One possibility was that each wasp, although it could not see its nest from a distance, was perceiving it directly with some other sense organ, perhaps one of an unknown nature. There was an easy way to check this. Empty biscuit tins without lids were dug into the path so that their rim was level with the ground. The soil in the tins was firmly stamped down so that it became as solid as the original soil. Sooner or later wasps started to dig nests in these tins. When one of these wasps was away hunting, the tin, containing her nest (tin A), was taken out and dug in about a yard away, while another identical tin (B) was put in its place. When the wasp returned she never went to her displaced nest in A, but always went to B and searched at exactly the spot where her nest 'ought' to have been. That showed that she could not have responded to any stimuli emanating from the nest itself. It rather seemed as if, like *Philanthus, Ammophila* used landmarks. If so, she must have learned them, for the landmark situation was naturally different for each burrow.

Some nests were not far from objects that could be potential landmarks, such as tufts of grass or pine cones or pebbles. When such objects were moved a foot or so, the returning wasps were misled; they looked for their nests in the correct position relative to the displaced landmarks, not in their actual position.

Next the Baerendses tried to find out how far afield the wasps' knowledge of landmarks extended. They caught wasps when they arrived at their nests, transported them in a little gauze cage wrapped in a black cloth and released them at different points in the heath. They then observed whether and, if so, how they got home. The results were very suggestive. First, all wasps that were released far from home—more than, roughly, 40 metres—wandered in irregular loops. Often such a disoriented wasp climbed a bush and then made straight for the path. The nearer home the wasps were released, the less likely was this disoriented phase to occur at all. It was, of course, significant that the limit was at 40 metres, beyond which the hunting wasps were only rarely found.

Many wasps were released in different directions on subsequent occasions. In all these tests it was clear that

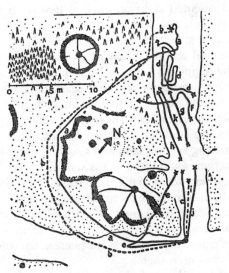

Fig. 11.   Eleven homing tests (a, b, c, etc.) with one Sand Wasp, released at crosses. The circle in lower right-hand corner indicates her nest.

there was an area round the nest from which they made their way home and that, beyond this area, they were disoriented. The size of this 'known' area was different for individual wasps and, what was more, the actual location of the different areas varied. Some wasps homed readily from a large area to the north-east, although they were lost even 10 metres from the path on the other side. Others had a known area in just the opposite direction. One wasp had a very remarkable record (Fig. 11). When transported over about 15 metres or more to the north-west, she returned home on a long roundabout way, starting off along the path away from the nest, turning left as soon as she had passed a large Pine tree, and travelling home in a semicircle round the tree. When released less than 15 metres in the same direction, she returned straight home. Yet in both sets of tests she appeared oriented right on release. It was obvious that this wasp knew the whole path and that only from the northern part she took the detour home—presumably because she had learned this part of the path on trips on which she happened to come from the heath beyond the tree. It rather suggests that they learn their way at first by retracing the route they followed in going out.

Another series of tests showed convincingly that the wasps did use landmarks all over the known area. Apart from replacing natural landmarks such as grass tufts that had been there before the wasps started to dig their burrows, the Baerendses made their wasps train themselves to artificial landmarks laid out after they had completed their nests. Much the same types of landmarks were used as had already been presented to *Philanthus*—blocks of various shapes and sizes, hemispheres, flat discs. The wasps always used them.

However, such small beacons were used only after the wasps had already arrived in the area near the nest. When they were moved more than about 3 ft., the wasps

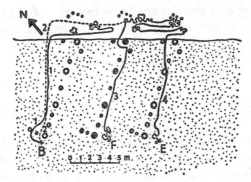

Fig. 12. Four homing tests with a Sand Wasp, nesting at open square, top center. Black circular dots indicate artificial trees in training position. Open circles, position during tests 1 and 2 (left) and 4 (right).

did not follow them, but searched round in the nest area itself. How did they find their way from the hunting field to this immediate vicinity of the nest? With *Philanthus*, which flies over such long distances, we had been unable to investigate this, but the slow-moving and more earthbound *Ammophila* offered a better opportunity.

A colony was found nesting on a path in a rather bare part of the heath. Here the Baerendses, after again having gone through the tedious labour of ascertaining the extent of the hunting area, put out a row of large beacons at right angles to the path. As beacons they used artificial trees, large green Pine twigs forced into the tops of iron tubes of varying lengths. After having left them in position for a week or so, which they imagined to be long enough to make the wasps use them, displacement tests were made. When a wasp arrived with a caterpillar at a nest not further than 2 metres from the nearest beacon, she was captured and released near the other end of the row. If this wasp showed signs of disorientation (thus

showing that she did not know this part of the heath),
she was, of course, no use for this test and she was taken
back to her nest. Those, however, that went straight
home were followed and their route recorded. Then they
were captured again and now the row of trees was
moved parallel to its own position. Then the wasp was
released at the far end of the line of beacons in their
new position.

Most of these wasps showed that they were guided by
the 'trees'; they arrived at the wrong spot, as far away
from their nests as the distance over which the 'trees' had
been moved. A control test, in which the trees had been
put back, was always taken immediately afterwards and
it always brought them home again. Some wasps, however, were not influenced by the tests at all; since there
was a row of real trees at the other side of the path,
which they faced all the way home, it seems probable
that they used these as landmarks. To check this by digging out and moving all those trees was, of course, impossible.

Many more tests were done, but it would carry me
too far to discuss them in detail. The overall result was
that *Ammophila* finds her way home by using landmarks
of many kinds, the position of which she has to learn,
just as *Philanthus* does. It is quite possible that apart
from these landmarks she uses the sun as a compass as
well, but the tests were not designed to check this.

Thus another astonishing aspect of *Ammophila's* behaviour had been revealed. The analysis of homing
showed what complicated things each wasp had to learn
in order to enable her to care for two or three nests
simultaneously; she has to associate her findings at the
'call', when she receives her instructions, with the landmark situation which will lead her to the particular nest
she has to supply. It is also worth considering how ac-

curately her behaviour has to function in order to guarantee the larva's survival. The slow egg-laying rate does not allow for many larvae to be sacrificed. How remarkable that such a system could develop, so different from that found in, for instance, moths and butterflies—which just lay their eggs and then abandon them.

It is impossible to devote much more space to the many other aspects of *Ammophila's* life which the Baerendses studied, such as the highly specialized way of paralysing the caterpillar, or the curious social resting and sleeping habits of the wasps.

One particularly interesting consequence of their work must not be omitted, however. As a result of the great accuracy of their work, the Baerendses ultimately discovered that they had been watching a wasp that had up till then not been recognized as a separate species. While they were engaged in this work, another observer, A. Adriaanse, was doing similar observations in another part of the Netherlands. At first neither party knew of the work of the other. When Baerends published a short account of his first findings in 1938, Adriaanse saw that his notes did not at all tally with Baerends's story. He found what might seem to be a triviality: his wasps, instead of blocking the nest entrance with material gathered from all over the environment of the nest, had a special 'quarry' near the entrance where the material was stored every time the nest was opened.

Many people would have considered this just a local difference, perhaps induced by a different environment, but Adriaanse went into it more carefully. And he found not only that there were more differences, but discovered a colony in which he could observe both types of behaviour—those observed by Baerends and those seen previously by himself. By marking the wasps with colour dots he discovered that some wasps behaved consistently in the way described by Baerends, others as seen by himself. He concluded that he had to do with two forms, consistently different in behaviour even when living in the same environment. Here are some of the differences between the A(driaanse) and the B(aerends) forms:

A	B
Provides larva with larvae of *Tenthrenids* (saw flies).	Provides larva with lepidopterous larvae.
Has only one nest at a time.	Has 'telescoped' broods.
Breeds from end of May till beginning of August.	Breeds from early June to middle of September.
Closes nest with one block, covers this with sand from 'quarry'.	Closes nest with many blocks, sand taken from all round nest.
When opening nest, puts sand back into quarry.	Carries sand away in flight and scatters it.

Apart from these there were other differences. Now before all this had been observed, another entomologist, Dr J. Wilcke, again working independently, had been studying the taxonomy of digger wasps and had found that museum specimens marked 'Ammophila campestris' were not, among themselves, as alike as could be expected. He thought there seemed to be two types, but being engaged on other work at the moment had shelved this little problem. The discrepancy between Baerends's and Adriaanse's work made him have another look at his wasps and he now discovered that *Ammophila campestris* must be considered a collective name for two distinct species. The morphological differences were minute, but they were consistent.

Adriaanse now settled the next question. Were the behaviour differences between 'A' and 'B' really correlated with the morphological differences as found by Wilcke? He watched a number of his wasps and collected a series of them, labelled them 'A' or 'B' according to their behaviour, then sent them to Wilcke. The correlation with the morphological work was perfect: all 'B' wasps were identical to the form Wilcke had baptised *Ammophila adriaansei*, all A wasps were *Ammophila campestris*. Next Adriaanse collected pairs actually found mating. Again it was found that the pairs were either both *adriaansei* or both *campestris*. In other words, although the two forms are so extremely similar, and even breed in mixed colonies, there is no inter-breeding between them; they are true species.

This whole story has various interesting aspects. First, it was rather pleasing to us behaviour students to see how behaviour characteristics could be used to help taxonomy. Second, since the behaviour showed so many differences even in wasps living in the same habitat, we had a beautiful demonstration of genetically determined differences in behaviour. Third, it raised evolutionary problems. Since the two wasps are so similar, they must be very closely related indeed. How did they ever diverge, develop a taste for different kinds of prey, make their nests in slightly different ways, etc? And why don't they interbreed? How much more discriminating are these wasps than were the zoologists who had never discovered that they were different at all!

The war interrupted this work. Adriaanse has unfortunately died, Baerends had to turn to other work. But perhaps *Ammophila* will one day find other biographers who will take up the threads where they were dropped. Or maybe this work will stimulate others to look into the life histories of other digger wasps. It is hard to believe that these two *Ammophilas* should be so much more interesting than other digger wasps; I prefer to think that each of the others will be found to be just as rewarding once it was studied with as much care and love as was *Ammophila*.

CHAPTER SEVEN

STUDIES OF CAMOUFLAGE

It may be commonplace to say that the variety of colour patterns found in nature is almost endless. Yet, in spite of the existence of numerous excellent books and illustrations, the majority of people, and even of zoologists, have only the vaguest idea of the true and immense extent of this variety. The naturalist and the artist are perhaps better acquainted with this aspect of nature than most other people. The naturalist, while sharing the artist's admiration of all these different patterns, is always wondering why they should be there at all. People have long ago abandoned the idea that they are there for our enjoyment; and the great naturalists of the past—Bates, Wallace, Darwin and Müller—have presented hypothetical explanations suggesting that many colour patterns are of great advantage to their bearers who have, in a sense, been forced to develop them, on penalty of extermination.

In the absence of experimental work, these ideas have remained hypotheses. In the last few decades, however, much experimental work has been done; in our summer camps in Hulshorst, and later in Oxford, we have also contributed something to the growing body of knowledge on this subject. Part of our work was concerned with camouflage, part with warning colours, and part with signal colours. Since I have written elsewhere[1] about the signal colours, I will not discuss these again.

When we started our observations one main question had already been settled. It had been claimed long ago that animals that are camouflaged to us must also be invisible to their natural predators, and thus would derive protection from camouflage under natural conditions. Critics, however, did not fail to point out—and correctly—that it was not at all proved that natural predators respond in the same way as we do. Indeed, as evidence about the different types of the sensory functions of animals began to pile up, it became clear that many predators respond to quite other aspects of their prey than their visual appearance—for instance, to their scent. As we have seen, *Ammophila* seems to be one of these and, as a consequence, succeeds in living entirely on perfectly camouflaged caterpillars. Over-critical zoologists have even swung to the other extreme and, instead of giving the camouflage hypothesis the benefit of the doubt, have claimed that it was all nonsense. It is astonishing to read some of this would-be critical writing, in particular those passages that sneered about the 'arm chair attitude' of those who believed in camouflage—astonishing because, by any standards, it was the criticism that came from the study; the believers were naturalists. However, this belongs to the past, or at least almost so. Some critics are still gallantly but stubbornly holding out, fighting what has by now become a hopeless battle. In a number of cases the camouflage hypothesis has been subjected to experimental tests and proven right.

One of the most convincing series of tests was done by Sumner. As a prey he used a fish, *Gambusia*, which, like so many animals, can change its colour to match its background. Once adapted, it needs several hours to

[1] For full references, see p. 292.

re-adapt itself when exposed to a new background. Sumner's tests consisted in taking equal numbers of dark and light fishes, admitting them into one large tank, which was either light or dark, and then subjecting the whole group to predation. When about half the fish had been taken, he stopped the test and counted the numbers left of each category. As predators he used predatory fish, penguins and herons. In all tests he found that more camouflaged fish survived than non-camouflaged. Based upon many hundreds of fish in all, and involving three different types of predators, Sumner's tests are very convincing.

Similar work has now been done with several other camouflaged species and it has become impossible to ignore or deny the conclusion that camouflage works in nature. But all studies published so far have been concerned with one aspect of camouflage only—that of the matching of the overall colour of an animal with its background. Camouflage, however, involves more than that. I will just mention a few examples, referring the reader to the beautiful book of Cott for a detailed review of a multitude of examples and of the principles involved.

Half-grown larvae of the Pine Hawk Moth, which was common in the Hulshorst Pine woods, have exactly the same colour as the pine needles on which they feed. But if they were a uniform green they would not be very difficult to see, since they are so much thicker than pine needles. However, they have a pattern in which green bands, running along the body, alternate with white bands. They thus appear to the eye 'disrupted', broken up into units which blend with the pine needles. This disruption is found in most camouflaged animals. A striking example is the pattern of irregular dark dots found on the light buff downy plumage of young gulls. Such dots, or in other animals stripes or bars, not only break up the body's outline, but may blend with similarly coloured and shaped objects or shades in the environment.

Another, quite different, principle is that of countershading. Rounded objects, however similar in colour to their environment, are shaded and thus may stand out visually as something solid. To many caterpillars this would be most revealing since they live among flat leaves.

Fig. 13. Third instar of a caterpillar of the Pine Hawk Moth.

But many of them are much darker on the side normally turned towards the light, the dark side grading out into a lighter hue on the other side. The effect of this 'counter-shading' is almost incredible; it can make the cylindrical caterpillar appear perfectly flat, and it is amazing how even quite large caterpillars manage in this way to disappear among the leaves. I have often shown such caterpillars to visitors and watched their incredulous expression as I turned a twig round (so that it was illuminated from the wrong side) and suddenly made them realize that they had completely overlooked a fat larva as big as their little finger.

Many animals have specialized on still another trick—they resemble objects that the predators do see, but that they ignore because they know they are inedible. The twig caterpillars are perfect examples of this 'special resemblance'. Here again colour resemblance is involved, but the effect of inconspicuousness is attained by much more than this.

Of all these principles only that of general colour resemblance had been experimented upon when we started our work; the others had merely been recognized as having effect on human beings. It seemed worth while to extend experimental work to predators' responses to these as well. The attractions of such work are manifold, because one has to study the predators' behaviour as well as the colours, shapes and also the behaviour of the camouflaged animals, for no camouflage works unless the animal has the inclination to select a suitable background and to keep still when in danger.

The abundance of insects in Hulshorst which possessed one or other of the alleged adaptations was too inviting to ignore and, consequently, as our field work expanded, we decided to raise Jays and Chaffinches and take them with us to Hulshorst where they shared camp life with us while being subjected, as suitable predators, to experiments. The pilot tests we did with them gave such interesting results that some of us decided to take up this work in earnest.

The unreliable weather often forced us to interrupt our studies of the digger wasps. During those intervals, which sometimes lasted weeks, we combed the forest, the heaths and the sands searching for insects. During many such trips we discovered a rich variety of forms, which apart from contributing to my students' and my own zoological education, provided us with numerous objects

for our work on camouflage. Counter-shading was found
in the larvae of the Eyed Hawk Moth (which was common on the Willow shrubs on the heath and the sands),
in those of the Kentish Glory, which we occasionally
found on Birch, in Lime Hawk Moth larvae, which also
lived on Birch and in the last instar of the Puss Moth.

The fauna of the lichens, which covered many of the
trees in the wood, provided us with lovely examples of
disruptive coloration. Several moths, a bug, a beetle and
a spider lived on them. All were of a light bluish-grey
ground colour, exactly matching that of some of the
lichens, and all had a pattern of irregularly shaped black
dots blending with the shades and dark patches of the
lichens. The Peppered Moth was one of them. It was
amazing how completely such a large moth, resting in the
open, could disappear from sight by mere trickery. On
Oak leaves we found a caterpillar which, in our fauna,
is perhaps the most specialized example of concealing
coloration, combining counter-shading with 'special resemblance'. This was a species that does not occur in
Britain and is rare in Holland—*Hoplitis milhouseri*, called,
quite unjustifiably, 'the Dragon'.

It was about an inch long, very fat and squat. Its
ground colour was a beautiful bluish green, exactly
matching the colour of Oak leaves. Here it was usually
hanging upside down and, in accordance with this position, its back was dark and the smooth grading of blue-
green into the yellowish-green colour of the under-side
gave it perfect counter-shading, making it appear as
flat as an Oak leaf. Its head, however, and, amazingly,
some sharply defined areas of the body surface were a
pale buff; and these areas matched precisely those small
dead patches found on most Oak leaves, showing where
a mining insect had eaten the leaf away except for the
cuticula. Of course, this mimicking of damaged leaves is
quite common in subtropical and tropical insects, but

few people have ever seen these forms in their natural habitats, while illustrations in books, which are often of mounted specimens, usually fail to demonstrate the full beauty of this kind of colour adaptation.

On Birch and Oak we also discovered numerous 'twig caterpillars', mainly those of the Peppered Moth, besides *Ennomos alniaria* and *Ennomos quercinaria*. Since it was with these masters of deception that we started our experiments, I will describe them in some more detail. But by pointing out some of the more striking features I think I can add life to the picture.

Quercinaria is the more striking of the three. Its colour is exactly that of second-year Oak twigs. Its shape is still more remarkable. It has the wrinkled texture of such twigs. Its two pairs of hind legs fit so smoothly round the twig that you cannot see where they end or the twig begins. The body is covered with most unexpected humps and ridges, again extremely similar to those of real twigs. Some of these ridges carry the likeness even further—they imitate in almost ridiculous detail the scars of last year's leaves, with the small dots indicating the severed vascular bundles. The thorax and head are no less remarkable. When the larva is at rest, it stands out rigidly from the twig on which it sits and then it folds the two foremost pairs of legs against the body, where they fuse with it, the head suggesting the end bud. But the third pair of legs stands out, giving an exact replica of a side bud. These larvae are further very clever in selecting twigs they resemble, in taking up the position seen in the photograph, and in keeping motionless most of the day, particularly when the bush on which they are sitting is slightly jarred. Everything, therefore—colour, shape, behaviour—fits together to make the animal perfectly camouflaged.

Since we found these larvae in considerable numbers (having overcome their camouflage by just shaking them out of small Birches growing on the bare sand, which made the larvae fall on a non-matching background), we decided to do some tests with our birds. There was a special reason for selecting these extreme examples of highly refined camouflage: if natural selection must be held responsible for these extremely detailed imitations of twigs, then it ought to be possible to find the selectors responsible for this. We would have to find at least one selector which would not only be deceived by twig caterpillars, but which could immediately detect slight defects of the adaptation and so keep up the pressure towards perfection. One objection that is often made against the theory of natural selection as applied to these phenomena is merely based on the alleged improbability of birds, or any animals, having sufficient discriminatory powers to act as such nice selectors. I for one, having learnt some very remarkable things about the vision of birds in the course of other work, had little doubt that they could be much more severe selectors than human beings, but until we could subject them to hard and fast experimental tests such convictions were not enough.

Jays were the first birds we tested; and a good choice they were, as will be seen later. Since we wanted to work with captive birds which had no experience of this type of insect food, a new aspect was added to our camp life. Months before we were due to leave for Hulshorst we had to take young Jays from their nests, raise them by hand and get them used to life in an aviary. Then we had to take them along to our camp and give them suitable living quarters there. We built four collapsible living cages, measuring $1 \times 1 \times 1$ metres, and put one bird in each. The cages were erected on the ground in the wood, under the shelter of a canvas roof. Adjoining them was one large experimental aviary, not covered by the roof. By small doors that could be opened and closed by us, each Jay could be admitted singly to the larger aviary, where it could then be watched when it came face to face with our set-up.

The very first tests, simple though they were, gave us some excellent results. They were done with a number of *Ennomos alniaria*, which were more abundant than *quercinaria*. In spite of their name, they were all taken from Birch and resembled Birch twigs. Some of them were offered to the Jays on leafy twigs, sitting in their natural position; others were laid out on the ground and, in addition, we gave some real twigs about the size of the larvae.

Our Jays had been trained to being fed in the large aviary. Usually they received mealworms and now and then they had been offered various wild insects, but they had never had any twig caterpillars. They had, however, seen and handled twigs. As captive Jays investigate everything that is new to them, we had to provide them with an opportunity to do this and so become used to twigs as their wild brothers are. Being admitted into the aviary had come to mean being fed and, consequently, every Jay, upon entering the aviary, began to search round. It could not know, however, what kind of food to expect; every Jay began the test completely 'open-minded'.

To our delight our first Jay failed to find the larvae at first. It hopped past and over several of them, giving an exploratory peck here and there. This went on for about 20 minutes. We were already thinking of what to do next, when suddenly the Jay stepped on a larva. This was too much even for an *Ennomos* and, giving up the advantage of keeping motionless, it began to wriggle. At once the Jay picked it up, beat it on the ground once or twice, and swallowed it. This gave us already an indication of the importance of stillness in combination with camouflage. But what happened next was a complete surprise to us. The Jay, immediately after swallowing the larva, looked round and hopping briskly through the cage, picked up —a twig. It nibbled at it, threw it away, hopped on, took another twig, and so on, one after the other, in quick succession. Then, after several of these mistakes it calmed down and ignored twigs and caterpillars alike.

Now this was one of those lucky extras that even rigidly planned tests sometimes provide quite unexpectedly. If the Jay had merely left the larvae alone, we could not have concluded very much. It could have been because it was satiated or disturbed. At best, a control test with larvae standing out on, say, a white background and in the absence of twigs, might, by giving a different result, have made the first test significant. But here we had a real positive result: the Jay had taken up twigs, but only after it had discovered a larva—which itself had obviously been an accident. Thus we had seen, in one test: (1) that the Jay did not see the larva until it moved, (2) that it would eat it eagerly once found, (3) that it confused twigs with caterpillars, and (4) that, therefore, it had originally ignored twigs and caterpillars alike because they were all 'just twigs' to it.

After this first encouraging result my fellow-worker L. de Ruiter, took up the work in earnest. Some of it was done in Hulshorst in 1946 and 1948, the rest in Oxford in 1951. He used in all seven Jays and two Chaffinches, and many caterpillars of *Biston strataria, Biston hirtaria* (Larvae of Peppered Moths) and, for the majority of tests, *Ennomos alniaria*. He could confirm our original conclusions, but had some extremely interesting additional results as well. It is worth quoting his findings:

'1. The first very striking point was that it took the Jays a long time to find their first caterpillar. They had always gone without food for at least 12 hours before the experiment . . . they always searched for food eagerly. . . . Yet it took the Jays in 6 experiments 34, 16, 29, 20, 7, and 40 minutes respectively to find their first caterpillar. . . .

'The significance of this will be clear when we mention that we had to stop experiments on camouflage in grasshoppers because the Jays always found each grasshopper within 10 seconds, however well we hid it. Evidently the Jays did not notice any difference between the caterpillars and the inedible sticks. . . . .

'2. The behaviour of the Jays was very interesting after they had accidentally found their first caterpillar (e.g., by treading on it). Often they immediately began to peck at sticks and caterpillars indiscriminately. This happened in 4 out of 8 experiments. . . . The change in the behaviour of the birds in these cases was especially striking after the complete absence of interest in the sticks in the first part of the experiments. . . .

'In two experiments, however, the Jays did not peck at the sticks after finding their first caterpillar. In two other experiments, although one or two twigs were taken, there was a noticeable difference in the behaviour of the bird towards sticks and caterpillars respectively. Sticks were taken hesitatingly and sometimes refused altogether, or taken only when all caterpillars had been eaten. Caterpillars were always taken very eagerly. Almost always practically all the caterpillars were found soon after the first one.

'3. In other words, some natural enemies of the stick caterpillars are able, at least at close quarters, to distinguish the caterpillars from sticks. If this is true the high degree of perfection with which the insects imitate the twigs of their particular food plant would seem to be of great importance. We tried to test this as follows.

'In 8 experiments, 4 with Jays and 4 with Chaffinches,

we gave, apart from twigs of their own food plant, sticks of different species as well—e.g., in the experiments with *Biston hirtaria,* which was found on *Prunus,* we used, apart from 2 types of Cherry (smooth and rough): Birch (brown), Dogwood (green), Bracken (yellow) and Sycamore (red leaf stalks). In these experiments the Jays confused the caterpillars only with the sticks of their own food plant, never with other sticks. . . .

'It seemed very remarkable that the survival value of such close harmony between insect and plant would be great enough to be shown in such a crude experiment. We therefore repeated the same test twice with Chaffinches. This gave essentially the same result.'

Thus it appeared that we had indeed had a lucky break with the choice of our birds; they had shown us that selectors of the required 'severity' actually exist.

Another part of De Ruiter's work concerned countershading. As I have already mentioned, the caterpillars of the Kentish Glory, those of the Eyed Hawk Moth, and the last instar of the Puss Moth were beautiful examples of this type of camouflage. The Eyed Hawk Moth combines a disruptive pattern with the counter-shading. A number of curiously coloured lines runs obliquely across the body. Each consists of a pale band, almost white, bordered by a dark zone, which fades into the larva's general body colour at its anterior edge. In the caterpillar's natural position on willow twigs, where it hangs upside down, the dark lines are under the white stripes and together they give an amazingly convincing 'fake' of three-dimensional ridges. Thus the larva not only appears flat as a result of its general counter-shading, but it is also broken up in smaller units by what might be called 'three-dimensional disruption'; and by these two devices it blends beautifully with its environment. Even full-grown larvae, as big as one's little finger, are often extremely difficult to find.

The Puss Moth larva combines counter-shading and disruptive coloration in another way. A white line runs along its side from head to tail, making a peculiar, incongruous sharp bend on its way. The caterpillar's natural position is again upside down, belly towards the light. This side is dark green; it fades towards the white line. The area on the back is also counter-shaded; it can be light bluish or purple. In the natural position both areas appear flat, but the back, although actually much lighter than the ventral side, appears much darker because it is in the shade. The white borderline separates the two areas visually, the outline of the caterpillar being disrupted. Again only those who have seen those creatures can appreciate how perfectly flat the animal appears.

The Puss Moth has still another string to its bow. If you touch it sharply, it will suddenly give up its camouflage and, with a jerk, turn its head and thorax towards you. The brown head is withdrawn between the 'shoulders', which bear a bright purple ring and two small black spots. The whole ring is rather like an ugly head with two small black eyes. This is probably a case of 'false warning colours' and comes into action only after a touch-stimulus has informed the larva that it has been discovered by a predator.

Does counter-shading do the same to the natural predators of the Puss Moth as it does to us? Does the caterpillar derive protection, from appearing flat instead of solid? Of course, it need not appear flat to a bird. Much depends on the way in which the natural predators distinguish between flat and solid objects. We ourselves use, unwittingly, more than one criterion of solidity; shading is only one of them. That we use it is clear from the fact that objects painted on a flat surface can appear solid to us. But everyone who has seen three-dimensional photographs will agree that there is a tremendous difference between them and the two-dimensional picture which has to rely mainly on shading. Counter-shading would not afford much protection if a Jay would judge solidity by binocular vision rather than by shading.

How could this be studied? It would be necessary to compare the vulnerability of these caterpillars under normal conditions and under conditions which deprived them of the benefits of counter-shading. This could be done in various ways: one could turn the caterpillars upside down, or illuminate them from below, or one could paint them a uniform green. De Ruiter decided to apply the first and the third method, because he thought that lighting the cage from underneath might upset the Jays and make any experiments invalid.

The tests required considerable preparation. First, a good supply of larvae was essential. We therefore bred them—as well as large numbers of other insects to feed the Jays in between tests—in small breeding cages kept in the camp. Further, the experiment had to be as good a replica of the natural situation as possible. This meant that dense vegetation had to be provided (lest the Jay would find even the normal larvae at once). In Hulshorst, the aviary was decorated with fresh twigs of the caterpillar's food plants before each test. Later, in Oxford, where an aviary measuring 9 × 9 × 2 metres was used, the natural shrubbery fenced in by the aviary provided an ideal environment. The density of larvae had to be kept low, so as not to make it too easy for the birds. Therefore, in each test only 4 larvae were used. Lastly, the movements of the caterpillars had to be eliminated. This was essential, since inverted larvae tended to regain their natural position and thus moved more than larvae would in the normal position. Therefore, all larvae were first killed with cyanide vapour and then fastened on to twigs.

In each test, caterpillars in the normal position and in the inverted position were offered in equal numbers, and the behaviour of the Jays watched. The Jays, trained to expect food, and always reasonably hungry when admitted, invariably began a thorough and systematic search, hopping from one branch to another and peering intently into the foliage, with extravagant contortions of their necks.

Because each test had to be arranged so carefully, relatively few tests could be done, but the results were quite satisfactory. Using caterpillars of five different species (Kentish Glory, Eyed Hawk Moth, Lime Hawk Moth, Privet Hawk Moth and Puss Moth) and three Jays, he found that, when equal numbers of inverted and normal larvae were offered, 167 inverted ones were found against only 107 of those in normal position. The three birds reacted differently (one was much better at finding normal caterpillars than the others) and the five species of insects showed a different degree of protection (Lime Hawk and Eyed Hawk were better protected than the others), but the general conclusion—viz., that turning counter-shaded caterpillars upside down increases their risk considerably—was safe.

It could be argued that, since inverted larvae were mounted on top of a twig while normal larvae were hanging under a twig, the latter were just concealed by the twig and that this and not their counter-shading was the reason why they were found less often. However, on 85 occasions when De Ruiter observed accurately how a Jay approached a prey, he found that it came from below in 51 cases, from the same level in 15 cases, and from above in only 19 cases. If anything, then, the normal larvae were more exposed than the inverted larvae.

The way in which they approached the two groups of larvae was also strikingly different. Inverted larvae were spotted from a great distance and eagerly approached. The normals were seen only at fairly close quarters.

A small series of tests was also done with 12 pairs of Puss Moth larvae that were painted a uniform green. Of these, 7 inverted and 9 normal larvae were found, and no differences in the Jays' behaviour as described above were observed.

All these tests, apart from confirming the idea that counter-shading affords protection from Jays, impressed us once more with the powers of discrimination possessed by these birds. In detecting Eyed Hawk Moth larvae in their normal position, they were far superior to any of us, and the very fact that they found 107 normal larvae in all tests together shows that, even with their present, almost perfect counter-shading, all the species studied by us must be under strong selection pressure from predators of this type.

Incidentally, on our collecting trips, when all of us always competed keenly for the largest bag, young children always won. I believe that the best explanation of this is that they could devote themselves much more single-mindedly to the job. Not only were we grown ups always thinking of the tests as the final aims of the collecting, but we were also more intently on the look-out for other things we knew we might find. In this kind of collecting single-mindedness is an invaluable asset, and the birds may, in this respect, be similar to children.

Having these Jays in the camp gave us ample opportunity to do occasional observations on their behaviour in general. One of their amusing traits was their tendency to imitate sounds. I used to wake up the camp between

6 and 7 a.m. by walking from tent to tent whistling the reveille and tapping each tent lightly with a stick. After some weeks, our Jays began to give wonderful imitations of the tune I used to whistle. One morning they started to perform at 4 a.m.—and two of my students woke up, dressed hurriedly and came out rubbing their eyes before they discovered what had happened.

One of our Jays, a female, fell in love with one of my students. Whenever he approached the aviary she would interrupt whatever she had been doing, hop on the perch nearest to him, raise her crown, neck and shoulder feathers, droop her wings, turn broadside to him, and produce the curious, rattling 'kraah' note one hears so often in spring gatherings of Jays (see D. Goodwin's descriptions, 1951). She distinguished this boy from all other people without the slightest hesitation, even spotting him when he approached concealed behind five or six others. Showing part of his face to her was sufficient to throw her into frenzies of courtship. We had to implore our friend not to come anywhere near the aviary when a test was on!

Once during some pilot tests with live Puss Moth larvae we made an extremely interesting incidental observation. A Jay discovered one of these big caterpillars and picked it up, whereupon the larva gave its warning display and also squirted the contents of a gland it has in front of its forelegs right into the Jay's face. He jumped into the air, releasing the caterpillar and violently snapping his bill (a sign of fright), and then went into a curious display: lowering his head, he spread his wings and held them as a shield in front of his body, so that the two blue fields were almost touching each other. Not knowing the Jay's general behaviour very well at that time, I thought that this must be a threat posture. My friend Derek Goodwin, however, who has studied Jays intensely for years, both in captivity and in the wild, assured me that neither in threat nor in courtship does the Jay ever display in this way. However, they adopt exactly that posture when 'anting'.

Fig. 14.   Female Jay displaying.

Many birds respond to ants, particularly to Red Wood Ants, in a peculiar way. They stand or crouch among them, pick them up and rub them through their plumage. Jays do not pick up ants, but they approach them most excitedly, spread and lower their wings, and make these rubbing movements. The ants soon crawl all over them when they start this. The function of this 'anting' is still a complete mystery, but it is a wide-spread and easily recognizable type of behaviour.

Now the interesting thing is that our Jay showed the 'anting' posture as a reaction to the Puss Moth larva just when the latter was squirting. As is well known, the substance secreted by the caterpillar's gland is formic acid, the poison ants use! It seems likely, therefore, that the Jay responded to this chemical stimulus. And this is interesting again because, in tests on colour vision in Jays I had done some years before in the laboratory, I had received strong indications that Jays had an acute sense of smell, by which they could smell mealworms they could neither see nor hear. I never followed this up, but in view of the uncertainty about the acuteness of the sense of smell in birds I am still hoping some day to do more systematic work on the chemical senses of Jays.

To return to counter-shading, De Ruiter did some supplementary work to find out whether Jays use shade to recognize the solidity of objects. This he did in the laboratory. Again the Jays were kept in living quarters separated from the experimental cage, to which they could be admitted by a door. In the experimental cage they were faced with two food cups. Over one of them a horizontal cylinder was attached, over the other a flat piece of cardboard of the same apparent size and colour. The Jays were first trained to go to the cup over which the flat cardboard was attached. This was done by providing food only in that cup. With such tests, one has to avoid training the Jays to a certain cup or to a cup in a certain position (e.g., the righthand one) and, therefore, the cups and their location were varied at random. The Jays learned this soon and, in 219 tests, they went to the flat model 188 times and 31 times to the other. In 51 tests when no food was provided in either cup (just to make sure that they were not responding to the food itself), they chose the flat model 48 times and the rounded one only 3 times. Next the Jays were offered the choice between a counter-shaded cylinder and the flat model. Rather unexpectedly, they still chose the flat model 19 out of 20 times. This in itself did not show, of course, that the counter-shading did not make the cylinder more like the flat model, but only that it was possible for the Jays to distinguish between them. In fact, De Ruiter found that it was quite beyond his artistic powers to counter-shade a cylinder as well as nature had painted these caterpillars.

When next the counter-shaded cylinder was given together with the normal cylinder, the Jay took the counter-shaded one 18 times and the normal cylinder only 3 times. But 16 times they just did not choose at all. When a plain flat model was given with a flat model shaded like a cylinder, the choices were 20 versus 1 for the plain model; plain flat versus normal cylinder gave 20 versus 0.

Now the birds were re-trained, so as to make them expect food under the cylinder. After this they preferred a shaded flat model to a plain flat one (14–0).

Together these tests seem to be open to no other interpretation than that shading made the flat cardboard look solid to the Jays, although the first test showed that the Jays did use other criteria of solidity as well—just as we do. Work done along similar lines with domestic fowl has given very similar results.

Although De Ruiter was handicapped in this work by a shortage of birds and a variety of other circumstances—all ultimately due to the limitations of funds available for such work—he has, I think, succeeded in demonstrating that counter-shading has survival value by offering considerable, though not perfect, protection against at least some predators.

De Ruiter then decided to have another look at the many different species of counter-shaded caterpillars to be found in our fauna. There were several reasons for doing so. First, it had always struck him that countershaded animals invariably take up a 'correct' position— that is, with their darkest side up. This position differs from one species to another. Many caterpillars, such as those he had used in his tests, hang upside down—and they are always darkest on the ventral side. Others, such as the Brimstone, turn their backs towards the light; and these have a dark back. The larva of *Apatura iris* normally hangs head down; and this larva is shaded from the tail, which is dark, to the head.

Counter-shading is also common in fishes. Most species, such as our Rudd and Roach, are dark on the back. But there is at least one fish, *Synodontus batensoda* living in the Nile and adjacent waters, which is darkest on its belly and white on the back; this fish actually swims upside down!

De Ruiter thought that it was worth investigating how these animals—which, of course, have not the slightest knowledge of the significance of counter-shading—managed to take up the correct position. A close study of this, and also of the way in which the beautiful gradation of the colour was effected, would be bound to give some insight into the complexity of the whole combination of properties that together result in concealment.

The second reason for extending such studies over many species was the following: it could be argued that the protective effect found in these tests was just an accident and not really an adaptation evolved in long periods of sustained selection pressure by predators. But if it could be demonstrated that all these different species attained the effect of blending with their environment by the use of many different devices, then it would become increasingly improbable that we would have to do with mere coincidences.

One could compare this with a non-human observer examining houses and wondering whether the slit found near the door, through which the postman slips his letters, just happens to be there and accidentally fulfils a useful function, or whether it may have been made there for just this purpose. If, by examining a hundred houses, he would find that each has one; further, that some have it in the door, others just beside it, others in a special box on the road side; and further, that some are cut out of the wood of the door, others are slits in a brass plate, others again are hewn out of stone; finally, that some had a wooden letter box behind it, others a glass-panelled box, others again a metal wire container, etc.—then he would certainly conclude that a letter box was an adaptation evolved for a very special purpose. And he would be right.

De Ruiter investigated the location and nature of the graded colour pattern in twelve species of caterpillars belonging to four different families. He also summarized the findings of Süffert on five more species belonging to three other families. He found that the origin and location of the elements that together make for countershading are widely different. The green and the shading can be due to the colour of the blood, to the skin or to the cuticle covering the skin, or to combinations of these elements.

He next studied the behaviour that makes the caterpillars take up the correct position. This again differed from one species to another. Some species orient themselves to light directly, while others respond to gravity—which, of course, since light usually comes from above, normally has the same effect. Other species orient themselves by using the structure and position of leaves and twigs; others again are just forced upside down by their own weight. And in some species such factors act in combination.

Thus all the evidence points to one conclusion: that the mechanisms responsible for this type of camouflage are very different in all these species, but that there is one, and only one, thing they have in common—their concealing effect. Taking this work as a whole, it becomes almost absurd to think of camouflage as anything but an adaptation, developed under selection pressure by extremely competent predators.

Yet all this work is only preparatory to research that aims at the heart of the problem. What one really wants to know is: how have these wonderful adaptations originated and evolved? This is, of course, a huge problem, with many aspects, among which the genetical aspect is prominent.

By a stroke of good fortune I have been associated, in a minor way, with an investigation into the genetical and evolutionary aspects of a case of camouflage. The Peppered Moth, *Biston betularia*, which as a larva is a beautiful stick caterpillar, is itself no less strikingly camouflaged. It flies only at dusk and by night, spending the day resting motionless on tree trunks, where it blends perfectly with the lichens.

About a century ago a moth was collected near Manchester that, although obviously a Peppered Moth, was almost uniformly black instead of chequered. This, appearing to be a melanistic mutant, was named *carbonaria*. Soon more were found and at present there are areas where almost the entire population is black, the pale, chequered form being exceedingly rare. These black populations occur in heavily industrialized areas, such as the Midlands and the London region. They are also found in the Ruhr area in Germany.

This has attracted the attention of geneticists and students of evolution for the obvious reason that here was a case of evolutionary change occurring under our very eyes. A few years ago, Dr E. B. Ford of Oxford University and Dr B. Kettlewell, who had both studied the matter in detail, decided to try to prove their hypothesis that natural selection had produced this change by favouring the black mutant in the industrial areas, thus allowing them to replace the original form.

Kettlewell decided to carry out a field experiment on a large scale. He bred huge numbers of both forms, collecting his stock in many different districts of the United Kingdom. By the time the pupae were about ready to hatch, he moved, with his caravan equipped as a field laboratory, into a wooded area of which the composition of the natural *Biston* population was known.

He selected two extremes for comparison—a wood in Dorset, where *carbonaria* was rare and the typical form abundant, and a wood in the Midlands, where practically all the Peppered Moths were *carbonaria*. In each wood he released, day after day, known numbers of both forms, never more than one on a tree. Each moth had been marked with a colour dot on the underside of the wing. After having left them on the trees for one day, he recaptured as many moths as he could. For this purpose he used traps which, as he could check, were equally attractive to both forms. All round his experimental area, which extended over several acres, he hung up muslin cages containing virgin females of both forms. When at dusk the moths began to fly, males were attracted to these females (for this reason only males were released) and could be captured while they were fluttering round the cages trying to get in. Kettlewell also used a mercury vapour lamp, the light of which the moths cannot resist. The captured moths belonged partly to the local wild population, partly to the population released by Kettlewell; the latter could, of course, be recognized by their colour dots.

This 'mark-release experiment' took several seasons of strenuous work. The results were convincing. In the Midlands wood 630 males were released in all: 137 typicals, 447 *carbonaria*, and 46 *insularia* (another mutant which does not concern us here). In all 770 moths were captured, of which 149 were marked. Of these, 18 were typicals, 123 *carbonarias* and 8 *insularias*. Of 447 *carbonarias*, therefore, 123 were recaptured, or 27.5%; of 137 typicals only 18 returned, or 13%. This must mean that the mortality, in one day, of the typicals was very much higher than that of the *carbonarias*. In the Dorset woods, where comparable numbers were released, the result was the opposite; here about three times as many typicals returned as *carbonarias*.

This result demonstrated that there must be a tremendous selection pressure and that it was very different in the two areas: in polluted woods the *carbonarias* were best off, in clean woods the typicals.

What could be the agents responsible for this selection? Dr Ford and Dr Kettlewell believed that it must be some predator. In clean woods, where the typicals were beautifully camouflaged, the *carbonarias*, resting on lichen-covered tree-trunks like the typicals, stood out conspicuously, at least for human observers. In polluted areas, however, there are no lichens; they have died off there, not being resistant to the factory fumes. Tree trunks without lichens are relatively dark; moreover, in polluted woods they are covered with a layer of soot. On these trunks the typicals had lost all their camouflage and it was almost pathetic to see how exposed they were —we could sometimes spot them from more than 50 yards away—whereas *carbonaria* blended very well with its sooty background.

Kettlewell, of course, suspected that some, as yet unknown, predator(s) hunting by eyesight took the conspicuous moths in both areas, thus causing the selection pressure that he had found. By elimination he had come to the conclusion that birds must be responsible. Here seemed to be a snag, however: neither ornithologists nor entomologists believed that any bird took such motionless moths on anything like the required scale. However, Kettlewell was not discouraged by this and patiently and doggedly set out to see for himself what happened. His results were striking: by merely watching his releases, he saw birds of several species taking large numbers of them. And one can imagine his delight when he found that they took them selectively according to expectation: mainly typicals in the Midlands, mainly *carbonaria* in Dorset.

When these results were published for the first time, in a short paragraph in Ford's book *Moths*, a reviewer, writing in an entomological journal, expressed his doubt about the validity of Kettlewell's facts. And this was where I became involved. Having made several films on bird and insect behaviour, I was approached by Kettlewell with the request to help by trying to take some convincing shots of various birds in the act of taking the moths. I was only too keen and thus had the chance, not only of seeing seven species of birds regularly preying on the moths, and of filming them at work, but also of joining Kettlewell in his field work.

The days were long and we used every minute of them. Early in the morning we went to the mercury vapour trap to collect recaptures and any other insect we were likely to use either in the day's tests or for the filming. We then put out a variety of insects on a tree to which we hoped to attract the birds to be filmed. After a short interval for breakfast, I was taken to my hide, a small canvas cubicle a couple of yards from the selected tree, from where I could cover it with 3 and 6 inch lenses.

While I spent long hours waiting for my chance, Kettlewell checked his recaptures collected from the trap and entered his notes in the logbook, then proceeded to colour-mark all moths taken from the 'hatchery' the evening before (the moths usually hatch at the end of the day). Then he made his round through the wood, releasing his marked moths, one on each tree. This done, he spent some hours inspecting and caring for the pupae in the hatchery.

In between he called at my hide, to see whether I needed any help. He then replenished the females in the traps. All this took the best part of the day. Usually we were only just through with camp chores, shopping and supper when it was time to switch on the mercury vapour lamp again, and then to start our rounds of the re-assembling sleeves, which had to be done from dusk to well into the night.

For the purpose of filming, we plastered a tree with

6 or 8 moths, and often with a number of moths of other species as well. By this we created an abnormally high density, but since we always provided equal numbers of typicals and *carbonaria*, the chances of being taken were the same for both. Also, unlike the real experiments, when one tree never received more than one moth, we were now interested mainly in getting the birds in front of the camera as quickly as possible.

Waiting in the hide for things to happen was never dull, even though many hours might go by before the first bird actually came to my particular tree. In the course of about a fortnight I was able to take six species altogether, some of which came so often that I had to give up filming them. Most of my films were made in the 'clean' wood, where I got Spotted Flycatcher, Yellow Hammer, Nuthatch, Song Thrush and Robin. In the polluted wood near Birmingham I filmed Redstarts.

When moths were released in a new part of the wood, there was often little or no predation for a day or two. Then, suddenly, a number of moths disappeared, always on trees close together. This happened now in one, now in another part of the wood. Obviously what happened was that one particular bird found one of the moths by accident and then proceeded to take more. I was lucky enough to see this kind of thing actually happen. I had just put up the hide in front of a tall Beech tree, about 50 yards from where the mercury vapour trap stood, and had begun my watch at about 6 a.m., just after we had emptied the trap. As usual, a number of insects that had been attracted by the trap, but had not been caught, had settled for the day on the surrounding trees. Soon after 6 a.m. I saw a Song Thrush approach the tree nearest the trap. It was feeding on the ground and had caught one small worm, which it carried in its beak—later I found that it had a nest with small young. When the thrush was about a yard from the tree, it suddenly saw a Bufftip sitting on the bark. It jumped up and snapped it up. Not satisfied yet with its catch, it went on foraging. But now it disregarded the ground, and hopped straight to the next tree, about 12 yards away. Here it looked up intently for some seconds and then hopped to the next tree to search this one. In this way it visited six trees in a row, then crossed to my side of the lane, making straight for my tree. Here it took two *carbonaria* and made off to its nest. From that moment this thrush was a regular visitor, during my filming sessions taking, apart from some Bufftips, 11 *carbonarias* and 4 typicals. The other scores made while I was filming were: Flycatcher 46 *carbonarias*, 3 typicals; Yellow Hammer 8 *carbonarias*, 0 typicals; Robin 12 *carbonarias*, 2 typicals; Nuthatch 22 *carbonarias*, 8 typicals. In the polluted wood the Redstarts took 12 *carbonarias* and 37 typicals.

It was quite interesting to see the birds actually at work and to notice the differences in their behaviour. Flycatchers and Yellow Hammers hovered persistently up and down the tree trunks, sometimes not more than a few inches from the bark. They did not miss a single *carbonaria*, but failed to find most of the typicals, which was particularly significant when they passed a typical moth several times. The Robin, watching the trunk from a vantage point on a bough or another trunk, flew straight to the moth it had seen from there. Twice I saw it almost step on a typical without noticing it. The Song Thrush looked up from the ground and then hopped up, sometimes as high as three yards. The Nuthatches were the only predators which often defeated the camouflage. Running up and down the trunks, they could see even the typicals stand out from the 'vertical horizon'. Yet even they usually took the black moths first and their total score was higher for *carbonarias* than for typicals, since they overlooked some of the typicals.

As is usual in field work, we could not help noticing innumerable things that just happened although we had not set out to see them. The mercury vapour trap was a treasure store. Hundreds, sometimes even thousands, of insects were trapped in it. It was fascinating to find the various Hawk Moths such as Privet, Poplar, Little Elephant and Eyed Hawk, the lovely Green Arches, Bufftips and many others sitting quietly and completely undamaged on the sheets of paper offered as shelter inside the trap. Once when we released a Privet Hawk it began to quiver and flew off in a flash, but it had not flown ten yards when it dropped like a stone into the long grass. When we went to see what had happened, we found it sitting on a dead twig with which it blended beautifully—a striking case of fast habitat selection completely fitting its particular type of concealing coloration.

On another occasion we released an Eyed Hawk on a tree, among the Peppered Moths. The Yellow Hammer appeared, took a black *betularia* and then picked up the Eyed Hawk. The Hawk Moth immediately displayed its bright 'eye spots' on the hind wings and the bird dropped it as if it were red-hot. The moth fell on the ground still displaying, the Yellow Hammer followed it, but did not dare touch it again; for several minutes it hopped round and round it in a state of great excitement and finally left it alone.

Another little extra which was both amusing and annoying was the interest shown by bats in the assembling traps. Many of them found out very soon that there was good hunting round the cages and we can only guess how many marked males they took before we could get hold of them.

The delights of field work are not merely biological. Near our camp in Dorset we found large clearings in the wood entirely covered by wild Strawberries bearing thousands and thousands of their lovely little fruits. If ever you stumble upon the same natural resource, collect a fair-sized bowl of them; it is the work of half an hour. Pour some cider, vermouth and gin on them, add plenty of sugar, and leave it standing for a couple of hours. Then you are ready to receive the most sophisticated guest.

The emphasis in these observations was very different from that in our studies of the homing of digger wasps. There we were mainly interested in what made the wasps behave as they did; here we were concerned with survival value, or function. In both cases our approach was analytical and I have often met people who were not sympathetic towards this approach. They argued that it made us forget the beauty of the things we analysed; they felt that we were tearing the wonders of Creation to pieces. This is an unfair accusation. We often felt that there is not less, and perhaps even more, beauty in the result of analysis than there is to be found in mere contemplation. So long as one does not, during analysis, lose sight of the animal as a whole, then beauty increases with increasing awareness of detail. It was typical that one of my friends who did not approve of our approach was himself a first-rate forester, who looked at his trees in a thoroughly analytical way—without losing sight of the beauty of the forest as a whole.

I believe that I myself am not at all insensitive to an animal's beauty, but I must stress that my aesthetic sense has been receiving even more satisfaction since I studied the function and significance of this beauty.

CHAPTER EIGHT

DEFENCE BY COLOUR

While we were busy with our studies of camouflage, we were of course, aware of the fact that many animals are not camouflaged at all, but have colour patterns that are conspicuous. A Mallard drake, a Kingfisher, and a Swallowtail Butterfly cannot be said to be camouflaged. The Peacock butterfly is beautifully cryptic (protectively concealing) when at rest, but when disturbed it either flies off, or just flaps its wings without leaving, and in both situations it is extremely conspicuous.

Of course, bright colours are not always incompatible with camouflage. A cryptic animal may be very brightly coloured, yet if its natural surroundings are equally colourful, it may be difficult to see. For instance, the bright yellow, black-banded shell of a *Cepaea* snail is anything but dull, yet, as the work by Cain and Sheppard has shown, it is camouflaged in certain habitats and derives considerable protection from its pattern. Even with this reservation, however, there is no doubt that many animals are the opposite of camouflaged—they are conspicuous even in their natural environment.

This was known to the old naturalists and various hypotheses have been put forward. But, as with camouflage, experimental work testing these hypotheses has not been done until quite recently, and when we started our work there was still much uncertainty. Now we know a good deal more. Before continuing my account of the contributions made by my co-workers, in whose studies I have taken part in one way or another, let me review the various theories that have been tested so far.

In some cases conspicuous colours have, long before Darwin, been considered to be effective in mating. Nuptial colours of many birds and other animals were considered to stimulate the females of their own kind and thus to be a means by which their co-operation in mating could be ensured. This idea has been expounded by Lorenz, who suggested that bright colours, just as sounds, postures and scents do, often function as 'releasers'; they serve to release appropriate responses in other animals and thus act as means of communication between individuals. Some of our own experimental work has been done with a view of testing this idea. The releaser function of a number of conspicuously coloured structures in such diverse animals as birds, reptiles, fish, crustaceans, insects and even molluscs has now been demonstrated by experiments.

Other instances of conspicuous coloration have been regarded as a means of defence against predators. These were thought to act in a variety of ways. (1) The yellow-and-black banding of wasps for instance was thought to advertise some unpleasant or repulsive quality and predatory birds were thought to learn to avoid them after a nasty experience. (2) Those in themselves harmless or perhaps palatable insects that had colour patterns similar to obnoxious ones—e.g., hover flies wearing a wasp-like pattern—were assumed to derive protection from this resemblance, the idea being that predators would, after experience with wasps, leave the hover flies alone as well. This is the hypothesis of mimicry in its strict sense. Further, (3) such forms as the Peacock butterfly, which normally are camouflaged but which suddenly display bright colours when attacked, were thought to be quite palatable but to scare would-be predators away by their display, as a kind of last resort when it had become clear that camouflage had not worked ('false warning colours'). And finally (4) some bright markings were thought to be 'deflection marks', which made the predator attack a non-vital part of the animal and thereby fail to kill it and even allow it to escape with minor injuries.

The exact formulation of these various hypotheses has not always been the same—different authors have had slightly different ideas—but it seems that at present experimental evidence has confirmed, not only the releaser idea, but also these four main hypotheses (true warning colours, mimicry, false warning colours and deflection marks) at least in those species which have been carefully studied. In addition, it is quite possible that there are cases of conspicuous coloration with still other functions—the most exhaustive review of these, as well as of the main four functions, has been given by Cott in his book *Adaptive Coloration in Animals*—but the evidence about such other functions is still circumstantial.

As I said before, these chapters will not deal with our work on releasers, but will be confined to anti-predator devices.

The work I have been associated with most closely is that done by A. D. Blest on so-called eye-spots in insects. Many insects have localized, brightly coloured patterns that resemble pictures of large vertebrate eyes. Such eye-spots have been described, for instance, in beetles, grasshoppers, moths and butterflies. Sometimes they are kept on show continuously, but usually they are found on normally concealed parts of camouflaged animals, being displayed suddenly when the insect is approached or actually touched by a predator such as a bird. In our own fauna the Peacock butterfly and the

Eyed Hawk Moth are typical examples. Beautiful eye-spots are also found in many Saturnids.

The Peacock rests with its wings vertical, the dorsal surfaces touching each other. In this position the camouflaged under-surfaces of the hind wings, and of part of the fore wings, are visible, and the brightly coloured dorsal surfaces are concealed. When such a butterfly is touched, or even when it sees a sudden movement nearby, it quickly opens its wings, then shuts them again. This is often done several times in quick succession. By these movements the eye-spots on the dorsal surfaces are suddenly exposed in a series of brief flashes. This display is made more effective by the butterfly turning and tilting its body in such a way that the broad surface of the wings is turned towards the source of disturbance. You can easily provoke the display and make the Peacock turn and tilt, by waving your hand near a resting animal on a chilly day (when it is warm it will just fly).

The Eyed Hawk Moth is a camouflaged species that rests by day, usually on the trunk of a tree, and flies at dusk. In the rest posture it has the fore wings folded back so that they cover the hind wings. Thus only the dorsal surface of the fore wings is visible. The display is best elicited by a slight jab with a hard and sharp object. The Moth spreads its wings, exposing the eye-spots situated on the dorsal surface of the hind wings and then moves the wings up and down in a series of slow, emphatic flaps. It then comes to rest with the wings still spread and, if left alone, it very slowly moves back into the resting position.

Thus, although the behaviour of these two insects is different in detail, they are both camouflaged when at rest and both manage to display eye-spots when disturbed. This uniformity and constancy of the effect (viz., display of the eye-spots) and, by contrast, the variety in detail and in 'machinery' is still more striking when more different animal forms are compared; and this, of course, together with the intricate and 'improbable' structure of the eye-spots themselves, leads one to assume that the eye-spots are adaptive features.

The eye-spots of the Peacock and the Eyed Hawk Moth, however striking they may be, are not even the most specialized ones. Several other species have eye-spots that imitate real eyes in still more detail: the rings are not concentric, the dark 'pupil' is placed slightly off-centre, the eye is shaded, suggesting solidity; lastly, a tiny white dot imitates a highlight, as seen in the real eye.

When David Blest decided to study these eye-spots, he was faced with the same general questions as De Ruiter in his work on camouflage. First of all he had to find out whether the natural predators of these moths and butterflies are really intimidated by these displays. Further, if this were so, he wanted to study the mechanics of the response and the evolutionary origin of the eye-spots; if he found that these differed widely in different species, this would substantiate the assumption that we have to do with convergences. Thirdly, he wanted to investigate whether the incredibly subtle imitations found in some species are actually more effective than slightly less perfect imitations. Only if this could be demonstrated could he assume that selection by predators was responsible for the refinement of the adaptations. Blest's problem, therefore, was very similar to that of De Ruiter on camouflage, however different his actual subject was.

The results of his studies were very remarkable indeed. A good many observers have reported incidental observations on the responses of various birds to eye-spot bearing insects. I have myself seen Yellow Hammers and Robins pick up Eyed Hawk Moths from trees and, in 3 out of 4 cases, these birds dropped the moths at once when they started to display, as described in the preceding chapter. In the 4th case I could not observe what happened. A Song Thrush, however, ate an Eyed Hawk Moth in spite of its display and it would perhaps seem questionable whether prey of this size is not a little too large for small song-birds anyway—irrespective of whether it has eye-spots or not.

In our Hulshorst camps we had already done some preliminary tests with Chaffinches in captivity. When we presented the birds with Eyed Hawk Moths, they searched round until they saw one. Then they gave it an exploratory peck and jumped back as if stung when the moth began to display. In one case the bird did return after some hesitation and demolished and ate the moth; in two other tests the moths were left alone and the birds avoided coming near them even though they were no longer displaying. On another occasion we presented a Peacock to one of our Jays; when the Jay gave it a light peck, it flapped its wings, scaring the Jay out of his wits; he jumped straight up into the air and hit the roof of the cage. Yet, after some minutes, he returned and ate it.

We also gave the Chaffinches two moths that we had deprived of the eye-spots by just brushing off the upper surface of the hind wings; both these moths were eaten. Their display was vigorous enough, but they had nothing to show but two bare, greyish brown wings. This showed, in one and the same test, (1) that Chaffinches will eat such large prey and (2) that the eye-spots can scare them off. We also offered Privet Hawk Moths, an even larger species, to Chaffinches and these were all taken without much ado. These tests were, of course, too few to prove anything, but they encouraged us to tackle the problem.

David Blest used Yellow Hammers, Chaffinches and

Great Tits for his experiments. He started by carefully studying the behaviour of two Yellow Hammers capturing large prey. These birds were hand-raised and fed on mealworms and other food, but they had been prevented from having any experience with moths or butterflies. When they were 16 weeks old they were confronted for the first time with flying Red Admirals. In order to eliminate possible effects of the bright colours, the red and white markings were removed. It was found that at the very first occasion both birds attacked, killed and ate the butterfly offered. On subsequent occasions the birds did not change their behaviour except that they soon learned to concentrate the pecks at the insect's body rather than on the wings, at which they aimed many pecks in the beginning.

After this Blest did a series of tests comparing the responses to butterflies with and without bright colours. In the first test series, 4 Yellow Hammers were used. In order to study the effect of a bright colour pattern as such, the butterfly used first was the Small Tortoiseshell, which has brightly coloured wings, but no eye-spots. Intact butterflies were given as well as such that had their dorsal colours wiped off. While the displays of the intact butterflies scared the birds off in 9 out of 28 encounters, the colourless butterflies did not succeed in driving the birds back in any of the 18 encounters. A similar result was obtained with two Reed Buntings. It was clear, therefore, that the display even of the colours of the small Tortoiseshell had some repelling effect.

Then the display of the Peacock was put to the test. First, each of eight inexperienced Yellow Hammers was given 12 intact butterflies and 12 of which the eye-spots were brushed off. These tests were spread out over four days, care being taken to present the intact butterflies first in one half of the tests and the spotless butterflies in the other half. Sparing my readers further technical detail I will just briefly describe the results. In six Buntings, the spotless butterflies elicited altogether 37 escape responses against 149 evoked by the intact butterflies. These six Buntings, however, rapidly came to ignore the display and after relatively few tests ate the butterflies without much hesitation. Even after a rest of 30 days without any experiment the Buntings had not returned to their initial state: they could no longer be frightened by the displays.

The two other Yellow Hammers, however, responded differently. They attacked their first butterfly, were scared off by its display, then became increasingly timid, and left both intact and spotless butterflies entirely alone in all later tests. Four hand-raised, inexperienced Great Tits behaved in the same way.

These results showed convincingly that the display of eye-spots by the Peacock has survival value; moreover, the figures were much more convincing than those found for the Tortoiseshells, although the tests were not strictly comparable.

Of course, such experiments cannot tell us exactly how effective the eye-spots are in nature. But they do suggest that birds such as the Yellow Hammers and Robins I observed which dropped Eyed Hawk Moths as soon as they displayed, and then entirely abandoned them, did this because of the eye-spots. Yet a series of complete tests with intact and spotless butterflies and moths in nature would be very much worth doing.

After these tests were finished, Blest tackled the next problem: does any set of bright spots or bars do, or is an arrangement of concentric rings better than any other pattern, and, further, are really detailed imitations of eyes, with shading and a white spot, still better than just concentric rings? If it could be shown that the responses of the birds corresponded with the degree of refinement found in the most perfect eye-spot patterns of insects, it would become impossible to argue that selection pressure exerted by the predators could not have produced these eye-spots.

In order to test this it would be necessary to compare the effectiveness of a series of different patterns. To this end Blest designed a little gadget that enabled him to take large numbers of model tests without actually having to breed large numbers of insects. It consisted of a little box, on top of which a dead mealworm was deposited. On each side of the mealworm there was a little semi-opaque window, under which a slide with the desired pattern was concealed. A little 5 volt bulb could be switched on under each window, and only then was the pattern visible. When this box was placed in an aviary, the bird would come down. Exactly at the moment when it tried to pick up the mealworm, the observer, hidden behind a screen, would switch on the lights. The reaction of the bird to this 'display' would then be scored. Of course, at the start, it could not be predicted whether this way of flashing a colour pattern without actual wing-movement would work at all, but fortunately it did and this paved the way for the following tests.

Prior to each test series, every bird was allowed to get used to the box and real tests, in which the warning colours were flashed on, were started only after the birds had collected two mealworms without showing signs of fear.

In scoring the response of the birds, the highest significance was attached, naturally, to observations of a bird being scared away altogether, in other words, of survival of the mealworm. But even if the mealworm were ultimately taken, a bird might hesitate much longer with one model than with another, and these signs of hesitation were also compared. Roughly, in each test, if the mealworm was not eaten, the pattern received a score of 2; if it was eaten at once, the pattern scored o; hesitant behaviour scored 1. Therefore, the higher the total score of all tests with a model, the more effective this model was as a warning device.

It would lead me too far to discuss the experimental technique and the results in detail; but the following total scores were obtained in tests with various birds.

Two parallel bars at each side of the mealworm were presented and compared with a ring on each side. In 17 tests six Chaffinches, trapped and subsequently used to life in an aviary, scored: Bars: 25—Ring: 71.

A cross and a ring were compared in the same way, in 19 tests with six hand-reared Yellow Hammers. Scores: Cross: 11—Ring: 65.

A smaller but thicker cross was offered against the ring. Five trapped Great Tits were subjected to 11 tests. Results: Cross: 3—Ring: 30.

Next, five hand-reared Great Tits were tested with the

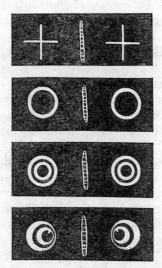

Fig. 15. Four patterns used in Blest's experiments.

larger cross and the ring, the number of tests being 14.
Score: Cross: 28—Ring: 44.

The same test was done with three hand-reared Chaffinches. The results of 7 tests were: Cross: 6—Ring:25.

Next, eight hand-reared Yellow Hammers were offered
the single ring and three concentrical rings. They scored
in 23 tests: Ring: 32—Three rings: 94.

Six hand-reared Yellow Hammers were then offered
two concentrical rings versus a 'shaded', 'glossy' and
'focusing' eye, 13 tests being done. The score was: Two
rings: 24—'Solid' eyes: 31.

The same models were then used in 12 tests with
seven trapped Great Tits. They scored: Two rings: 4—
'Solid' eyes: 42.

These dry little tables tell an extremely interesting
story. First of all, they show convincingly that rings
are more effective as a scaring device than parallel bars
or crosses. This was shown for three species of song-
birds. Next, a number of concentrical rings are more
effective than single rings. And, most remarkable of
all, the models which imitated solidity by shading, high-
lighting, etc., in the same way as some of the most intricate eye-spots actually do, scored higher than even
double rings. There can be little doubt, therefore, that
these highly specialized eye-spots are better warning colours than mere rings; and further that some birds at least
could act as selectors of such perfect imitations.

Another interesting aspect was the fact that even
hand-reared, inexperienced birds are intimidated much
more by rings than by the other patterns, suggesting that
the fear of eye-spots need not be learnt. This is in striking contrast to the way 'true warning colours' act; as
I will show presently, such colours, which advertise un-
palatability, are not avoided by birds until they have
eaten one or more prey animals with these colours.

Finally, it is of considerable interest that, although eye-spots have a stronger intimidating effect than bars and crosses, the latter do have some effect. This suggests that if an insect, by some mutation, acquired just any bright spot on the wings which would be displayed whenever it would open its wings (e.g., as a preparation to flight), this would already be of some benefit to it, and from then on selection would favour any further development in the direction of a ring.

Now people not acquainted with the wide-spread occurrence of such eye-spots in insects might argue that the rings on, say, the wings of the Peacock butterfly are 'just a freak', and that it is something of a rare accident, almost to be expected to occur in one out of more than a million animal species, considering the huge variation in structure found in the animal kingdom. There are, however, a great many species which have these eye-spots in one form or another and this alone renders the 'freak' theory improbable. In the Lepidoptera alone there are seven different major groups in which some species have eye-spots, and in each group there are many species which do not have them. A German zoologist, F. Süffert, comparing the colour patterns of many butterflies and moths, found that they can all be considered variations of one general ground-plan, a scheme of which is given in Fig. 16. Comparison of the different species bearing eye-spots shows that eye-spots have been derived from different parts of the wing. The eye-spots found on the hind wings of so many Saturnids (relatives of our Emperor Moth) are modified discoidal spots; those of the Peacock are formed by an ocellus and parts of the inner and outer band of the ocellus system; whereas the eye-spots of the Eyed Hawk Moth are just modified ocelli. Even within one genus, eye-spots have developed independently in several species: there are species of the genus *Presis* whose eye-spots have each been derived

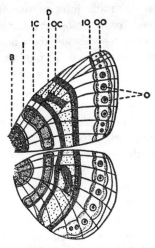

Fig. 16. Ground plan of the wing pattern of a Nymphalid butterfly, after Süffert.

D: *discoidal patch*	OC: *outer central band*
O: *ocelli*	IC: *inner central band*
OO: *outer ocellar band*	I: *inner patch*
IO: *inner ocellar band*	B: *basal patch*

from one ocellus, and others in which two ocelli have combined into one.

Therefore comparison suggests that the eye-spots of our present-day species have developed convergently and are true adaptations. This is confirmed by experiments concerning their function; these demonstrate at the same time the nature of this function—confirming the ideas of the earlier naturalists.

The intricate and 'improbable' nature of these adaptations becomes even more impressive when one studies the behaviour by which the eye-spots are displayed when the need arises. In most species the spots are concealed under a camouflaged coat, being shown only when some stimulus informs the animal that the protection derived from camouflage has not worked—as it is bound to on some occasions, since no degree of camouflage is completely effective. To begin with, the stimuli used may vary: we have already seen that the Peacock often responds to visual stimuli, but that the Eyed Hawk Moth has to be prodded. The movements themselves are also different. Blest has studied these in considerable detail. In all cases, the particular movement seen is suited to the structure extremely well; that is to say, they always result in the eye-spots being shown off. Species without eye-spots do not show these same displays. Totally camouflaged species, for instance, just stay put and keep motionless when prodded. Among the Saturnids there are species without eye-spots but with a bright pattern of light and dark bands across the abdomen, which are exposed when the moth, upon disturbance, falls, lifts its wings and curves its abdomen. In this posture the moth lies motionless for some time and it is then reminiscent of some large wasp.

The finer details of these correlations between displays and coloration are too complicated to describe here but they are given in full detail in Blest's paper.

While Blest's study seems to leave little doubt as to the real function of eye-spots, it raises one new problem. If it is of advantage to the insects to scare birds away, then it is disadvantageous to these birds to be scared of eye-spots. Why have not these birds been able to get rid of this harmful response? There seem to be two possible answers to this. One is that the organization of a bird's vision is so formed that a round pattern such as an eye-spot is more conspicuous than any other pattern, and that it is of benefit to the birds to be scared of anything conspicuous that appears suddenly. Another is that the escape responses of these birds are adapted in such a way

that they are released by sign stimuli which characterize
their own predators. If such birds recognize Owls, Cats
and Stoats, etc., partly by their eyes, it would be impossible for them to give this up in order to gain the
small advantage of an occasional extra snack. This second idea has received support from experiments by
Hinde, which show that Chaffinches do respond to the
eyes of Owls. However, the selection of the Owl's eyes
could be based on the same principle as that of the insects' eye-spots: they might be feared merely because
they are conspicuous. The crucial test would have to be
done with song-birds which do not respond to Owls at
all. If the second hypothesis were true, such birds should
not be scared by eye-spot displays. My guess is that this
second hypothesis is correct, and that all these insects so
to speak 'parasitize' on the song-birds' responsiveness to
their own predators.

So far, I have been discussing one type of eye-spot only
—the large, colourful and complex type. However, many
insects possess small and less elaborate eye-spots. A typical example is the Grayling butterfly. When at rest, this
species is beautifully camouflaged. When slightly disturbed, it raises its fore wings just enough to show a
circular black dot, with a tiny white speck in the
centre. Such small eye-spots, fairly common in insects
and other animals, have long been considered deflection
marks. The idea is that some predators, aiming at a
prey's head, would aim somewhere near the spot and
if the spot is located far from the head—as it often is—
it would allow the prey to escape with minor injuries.
There is circumstantial evidence that this might be correct. For instance, Swynnerton marked African butterflies
of the genus *Charaxes* with artificial eyes, lines, etc. near
the wing margins, released them, and recaptured them
at intervals. Marked butterflies tended to survive for

longer periods than unmarked ones and those that bore signs of having been attacked by birds showed beak marks and mutilations near the markings.

These results are suggestive, but not entirely conclusive. Sometimes the behaviour gives a clue as to the probable significance of an eye-spot. Cott describes how a fish, *Chaetodon capistratus*, which has an eye-spot near the tail, swims slowly tail first, but dashes off head first when attacked by a predator. This makes sense only if we assume that predators mistake the tail for the head, and so miscalculate the fish's movements.

Blest studied the problem of deflection spots by model tests. He fed mealworms to four hand-reared Yellow Hammers and found that the insects were usually attacked by pecks aimed at either the head or the tail, with a slight bias towards the head. Alternating with normal mealworms the birds were offered: (1) mealworms with either head or tail covered with white enamel paint with, in the centre, a small black circular dot; and (2) controls, i.e. mealworms with paint on one end of a colour closely resembling their own colour. The tests were taken in carefully planned sequence, so that it was possible to compare 'head bias' (preference for head over tail) with 'spot bias' (preference for spot over unmarked end, irrespective of whether the spot was on head or tail). The results confirmed the deflection hypothesis: whereas with unmarked mealworms the percentage of pecks directed at the head was about 60, the percentage of pecks aimed at the marks of those with artificial eye-spots was between 70 and 80. The percentage aimed at the marks of the controls was not different from the 'head bias' of the normal mealworms.

This problem was not pursued further, but it would be interesting to see whether animals with 'deflection eye-spots' would show the same extent of convergence as those with 'intimidation eye-spots'. Further, it might be expected that the optimum type of deflection eye-spot would differ from the optimum intimidation eye-spot. If so, one would expect to find two distinct types of eye-spots; although it is, of course, possible that there are markings which serve both functions and which, therefore, might compromise between the two ideal types. The whole problem is complicated still further by the possibility that some of these markings are used as 'intra-specific' signals as well, such as threat or courtship signals. My own guess is that they are in some species, such as the Peacock. However this may be, it is obvious that there is still much to be learnt about eye-spots and also that studies of this kind can have a bearing on more general problems of biology.

Other types of conspicuous coloration have also been studied experimentally, though not to any extent by myself or my associates. We have seen that some song-birds at least are scared by eye-spots before they could have learned anything about them. Their reactions seem to be released by the suddenness with which the eye-spots are flashed at them. But many species, such as wasps, show their bright colours permanently and there is, therefore, no such surprise element in whatever effect their colours may have.

The problem of whether, and if so why, birds avoid such yellow-and-black prey was tackled seriously by Windecker in Germany. He used the Cinnabar moth, *Euchelia (Hypocrita) jacobeae*, the black-and-red moth that can be seen flying in May and early June in areas where Ragwort (especially *Senecio jacobea*) is common. The caterpillars live on this Ragwort. They are extremely conspicuous when half or full grown, with alternating black and yellow rings, and they live in groups, often defoliating entire plants. Windecker showed that young birds do not hesitate to attack these caterpillars, but that when they take them into their mouth, they reject them with signs of disgust, such as violently wiping their beak. After that they refused them altogether. Windecker applied a simple method to find out what part of a caterpillar is distasteful: he mixed various components (insides, skin) separately with meal-worms and offered these to the birds. He found that mixing finely ground skin with mealworms spoilt them for the birds. But they did not object to the insides. He then shaved a great number of caterpillars and found that the hairs rather than the skin were responsible. In rejecting the caterpillars after having learnt that they were unpalatable, the birds did not respond to the hairs, but to the colour pattern, for they refused from then on any insect with a similar black-and-yellow pattern.

Comparable results were obtained by Mostler with wasps. Here it is partly the sting, partly the taste of the internal organs of the abdomen that offends the birds. Again, it took very few experiences for many birds to learn to leave wasps alone.

It seems, therefore, that such permanent, 'non-flashable' colours differ in their effect from the eye-spots. They are not effective against young birds; the species having these colours must pay a certain 'tax' to educate every new generation of their predators. They can only do this if they are in some way unpalatable—this being, so to speak, their true defence; and colour has come in on top of this. Insects bearing eye-spots are on the whole quite edible. Blest has found that the effect of the eye-spots tends to wear off, birds that face them often will gradually ignore them. With the true warning colours this is not likely. Firstly, Mostler found that a bird's memory for this kind of thing is amazingly good: one Redstart refused a wasp eight months after it had last met one. Even if there were some forgetting, renewed experience would soon refresh the bird's memory.

From these studies to a test of the mimicry theory was but a small step. If birds really learn to avoid unpalatable or otherwise repugnant insects by responding to their colours, then it seems probable that they will also reject other insects with similar colours. And actually such insects which, while themselves quite edible, imitate or 'mimic' insects with warning colours, are quite numerous. Many examples have been described since Bates first called attention to mimicry and the resemblance often involves much more than colour. In the books of Cott, of Hale Carpenter and Ford, and of Poulton many striking cases are described in great detail.

The best experimental investigation has been made by Mostler. He found that young song-birds are fond of various hover flies, even of those closely resembling wasps, bees and bumblebees. But once they have had their first experience with, say, a Honey Bee, they refuse not only bees but their mimics as well.

Mühlmann studied mimicry in tests with models. He painted mealworms in various patterns of red bands, made them distasteful with Tartar emetic (a substance causing nausea and vomiting) and, after birds had learned to refuse such models, recorded how similar to these a 'mimic' (a mealworm painted in a more or less similar pattern) had to be in order to be safe as well. He found that, while a pattern identical to that of the model had the best effect, even a quite superficial resemblance was enough to offer some protection. These results are of great interest because they not merely lend strong support to the mimicry theory, but show that even remote resemblance to an unpalatable insect has some survival value and could therefore be the starting point from which selection can act to produce more and more perfect mimicry.

Thus these various researches fully confirmed the

ideas of the earlier naturalists. Apart from this they also gave more detailed, and even quite new, results. For instance, as my colleague Dr P. Sheppard pointed out to me, the fact that one of Mostler's Redstarts refused wasps as long as eight months after his most recent experience with them, indicates that mimics must not necessarily be (as is usually assumed) less numerous than their models. The idea behind this notion was that if predators would have more experience with palatable mimics than with unpalatable models, they might learn to catch the models instead of avoiding the mimics. Mostler's observations suggest that this may depend on the degree of repulsiveness of the models. Who knows how many mimics a Redstart would reject on the basis of one or very few experiences with wasps? As long as the mimic's colour keeps repelling an experienced predator before it will try it, there would simply be no chance for it to learn that the mimics were edible.

Of course, colours may have still other functions in other animals; some types of coloration may even have nothing to do with the responses of other animals to them. The black colours of some desert animals, for instance, may help their bearers to radiate and thus lose heat as long as they keep in the shade—as many of them do.

What has always fascinated me about the kinds of colours I have discussed is that we have to do with *mutual* relationships between animals—both the coloured species and the animals responding to them show interesting behaviour—and it adds to the attraction of the work when one is faced with such surprising things as a predator being scared by the sudden display of eye-spots, or one undergoing the remarkable, though by no means unique, experience of discovering that a seemingly attractive bit of food is extremely obnoxious when tasted—an experience which changes its outlook on life profoundly. That our probing into these things is so easily rewarded when we do some simple research, and leads to an increasing insight into the history of living beings, is satisfying and encouraging to people who are foolish enough to spend their lives being curious about such trivialities.

CHAPTER NINE

BARK WITH WINGS

One of the commonest butterflies of the Hulshorst sands, and yet one of the least conspicuous, was the Grayling *Eumenis* (=*Satyrus*) *semele*. My first encounter with it made a deep impression upon me because of its very nearly perfect camouflage. We were watching the various insects gathering on a 'bleeding' Birch. Many Birches —and Oaks—in this region were attacked by Goat Moth larvae, which bored into the live wood. Through the wounds made by these large caterpillars the sap flowed out, its penetrating odour attracting numerous insect connoisseurs—Red Admirals, Camberwell Beauties, Tortoiseshells, Peacocks, Red Wood Ants, flies of many kinds, various wasps, beetles, etc. Other visitors were not interested in the Birch wine itself, but rather in some of the feasting guests. Wasps (*Vespa media* and *V. crabro*) came in numbers to catch flies and even the large butterflies; fly-hunting digger wasps (*Mellinus*) not only caught their flies here, but dug their burrows near the foot of such trees so as to have supplies at hand —building their homes near the supermarket!

While we were watching a Camberwell Beauty approach the tree—which it did as usual, flying against the wind, no doubt attracted by the scent—a small part of the tree's bark detached itself from the tree, shot through the air towards the butterfly, whirled round it for a few seconds, then abandoned it again, and dropped to the ground, where it vanished as suddenly as it had appeared. Cautiously approaching the spot where I had seen it hit the ground, I failed to see anything until a small piece of dirt leapt into the air and dashed past me, back to the tree, where it disappeared at once upon reaching the bark. This was a Grayling Butterfly, or 'bark with wings' as we called it.

Its camouflage was absolutely perplexing. When at rest, its wings were folded in the usual butterfly fashion, fore wings lowered between the hind wings, so that only the under surface of the hind wings showed. When seen from more than a few inches, this under surface was a very inconspicuous grey, mottled and striped in such a way as to suggest the irregular pattern of the bark of a tree. Yet when you looked at these wings through a lens, you discovered that the seemingly colourless pattern was made up of scales which were themselves beautifully coloured, either very dark brown, or white, or a golden yellow. They were arranged in such a way that there were no large fields of either one colour, the total impression being grey. These same colours were used on the dorsal surface of the hind wing and on both surfaces of the fore wings, but here they were arranged in large, more uniform fields. These dorsal surfaces were brown, with broad ochre bands, in which there were black circular dots with white centres, two on each wing. This pattern rendered the dorsal surfaces very colourful and conspicuous. The bright colours were visible in flight, but disappeared at once when the animal alighted.

The camouflage effect, and the means by which it was attained, were worth observing closely. The tree bark pattern was not just confined to the under surface of the hind wings, but was also found on those small parts of the fore wings' tips that were visible when the wings were folded; their boundaries coincided exactly with the edge of the hind wings—not a square millimetre of bright colour was shown when the butterfly was resting; neither was the bark pattern found on any part that was covered in the resting position.

The Grayling's movements showed the same specializations that many other camouflaged animals have: they either flew, and flew fast, or they were completely motionless. The change from dashing flight to complete stillness was so incredibly abrupt that we lost sight of a Grayling at once when it alighted.

We ourselves never did any tests to check whether colour and behaviour protect Graylings from their natural predators, but judging from experiments such as I described in Chapter Seven we can safely assume that this is so. Lorenz has described the way his tame Jackdaws responded to grasshoppers which behaved in this way. He reports how they were alerted time and again by jumping grasshoppers and rushed after them, but were rarely quick enough to catch them before they alighted and almost invariably lost sight of them after that.

The Grayling had still another kind of visual protection up its sleeve. When we approached a resting butterfly cautiously, it did not fly away at once, but it first lifted its fore wings a few millimetres, and waited in this position. If we did not move, it would stay where it was and after some time the wings would suddenly shoot back into their camouflaged position. But if we approached still more, it would fly off. The interesting thing was that the wings were raised just so far as to display the most conspicuous of the eye-spots—that on the tip of the undersurface of the fore wing. This looked like a textbook example of a deflection eye-spot; and now, after Dr Blest's experiments, I think there is little doubt that it acts as such.

Fascinating though the Graylings were, we did not pay much attention to them as long as we were busy with our observations on digger wasps. But when, in the late 1930s, *Philanthus* began to decline in numbers and further large-scale homing experiments became impossible, we turned our attention to other insects and soon saw that the Grayling offered good opportunities for study. We remembered that we had seen them fly after other butterflies and wondered whether this occurred regularly, and if so, why.

Fig. 17. Grayling showing its 'deflection marks.'

A few days' observations near a bleeding tree gave us the answer and made us aware of some very promising problems for field study.

We found that on sunny days the Grayling males (which were smaller than the females, had a paler band on the camouflaged wings and were a duller brown on the upper surface of the wings) took up regular observation posts, either on bleeding trees or elsewhere, but always on or very little above the ground. They usually sat completely still, but whenever another butterfly flew over, they would at once dash after it. Often, upon reaching it, they turned back at once, but when the other insect was a female Grayling, they followed it in a wild pursuit. Some of these females flew off as fast as they could. Others alighted, and then the pursuing male would alight as well and walk towards her. Thereupon, the female either walked away with a vigorous flapping of the wings or it sat motionless. Wing-flapping females were abandoned by the males, but when a female settled down and folded her wings, the male began to perform a fascinating, very complicated display. After alighting near the female, he walked round until he faced her. Then, with curiously jerky movements, he raised his fore wings step by step, quivering them while doing so, and with one quick forward jolt hit the female with them. All the time the wings were kept almost or completely folded. After this, he withdrew the fore wings a little and began to open and close their front edges in quick alternation, a very strange movement, since the rest of the wings remained closed.

While the male was doing this, he kept his feelers at right angles to his body, moving them rhythmically so that the ends described a circle or a semicircle every half second. The female stretched her antennae towards the male's wings. After a few seconds of this wing-fanning the male suddenly opened his wings, moved them slowly forward and, catching the female's antennae between them, pressed his wings together slowly but firmly. After this he walked round the female until he was standing behind her and a little aside and, bending his abdomen sideways, he made contact with her genital organs and copulated.

This was the full course of mating, but we saw it only a few times in nature. Most attempts by the males met with refusals from the females; as we saw later this was because most, if not all, females mated only once and the males had no means of distinguishing, from afar, between virgin females and those that had already mated; they flew up in response to all females and found out only later whether a female would respond or not.

It was clear, therefore, that the sudden flight of a male Grayling in pursuit of either a female or a butterfly of a strange species was the first part of his mating behaviour. It looked as if the sight of any butterfly stimulated the males to start this sexual pursuit and as if stimuli received after he had approached the other insect made him recognize whether it was a female of his own species or something else. The whole ceremonial display seemed to be a chain of separate responses. At that time we were particularly interested in the analysis of stimulus situations eliciting single responses, and here seemed to be a good object for such a study.

We—that is, B. J. D. Meeuse, L. K. Boerema, the late W. W. Varossieau and myself—began by studying the sexual pursuit a little more closely. It was soon clear that this was a response to a visual stimulus situation of a very unspecific kind, for the males followed a most curious variety of objects. We observed them pursuing 25 different species of Lepidoptera, among them such different types as Camberwell Beauties, Silver Y and Oak Egger. Many other types of insects were also followed— Dung Beetles, wasps, dragonflies of various kinds, grasshoppers such as *Stenobothrus* and *Oedipoda,* and even birds such as Great Tits, Chaffinches and Mistle Thrushes. These observations already indicated that the effective stimuli must be visual ones and this was clinched by the rather amusing cases of males flying at their own shadow.

We decided to follow up this lead and to expand the natural experiments we had seen by more systematic and planned tests. We made paper models of butterflies, tied them with thin thread to sticks of about 3 ft. long and with these 'fishing rods' presented them to resting males, making each model 'fly' towards a male from about 2 yards and making it pass at about a foot's distance. The males responded readily to these models and thus began a study which kept us busy—and our friends amused—through several seasons. The three or four of us would continually patrol miles of Grayling country and present our models in a more or less standardized way to every male we found. It would have been best to do a constant number of tests, using a constant number of models with each individual male, but this was absolutely impossible in the field, for males would often fly off after one or a few tests and could not always be followed long enough to subject them to all the tests we wanted to do. If possible, we always presented each model three times in quick succession (5–10 seconds interval). We then waited about 20 seconds and presented a second model in the same way. We went on with this, showing one model after another, until we lost the male. The length of a series, therefore, was determined by our males; they varied from 6 presentations to several hundred. We varied the sequence of the models irregularly (not being sophisticated enough to vary them in a random way) and also exchanged models every hour or so, so as to eliminate possible individual differences of angling technique.

All in all, we did some 50,000 tests. For each model, positive responses and number of presentations were recorded, and the number of positive responses per hundred presentations was taken as a measure of the releasing value of a model. Comparisons were made only between models of one series, for series done on two different days were, of course, incomparable because of changes in weather and other possible variables beyond our control.

We presented a remarkable spectacle (to say the least) in our practical but scarcely attractive field dress of shorts, broad-brimmed straw hats and sun glasses, each of us with his two rods, all with dangling paper models (some of which were brightly coloured), trying not to get them entangled in bushes or Heather, staring intently after one of our Graylings, trying to follow it on its erratic flight, running after it, suddenly stopping, stalking it, then carefully going through our angling ritual repeated three times, and finally making a few notes; all this with tense, serious faces. No wonder that we drew curious and suspicious glances from the occasional crofter or tourist who happened to see us at work.

Our first series was designed to find out whether the males responded to the details of the female's colour pattern. We had already seen that they would willingly follow butterflies as unlike Graylings as Red Admirals, but what we wanted to know was whether an abnormally coloured model was exactly as effective as a normal *Eumenis* female. In this first series we began by offering three models in succession—one plain greyish brown, cut out of coarse packing paper; another painted, with water colours, with as exact a replica of the real pattern as possible; and a third one on which real wings of female Graylings were glued. In another series we added a plain paper model covered with the same glue.

With these models we did over 2,000 tests and scored over 1,500 positive responses. None of the models received more responses than any of the others and we were certain, therefore, that the details of the colour pattern were quite irrelevant to the males, at least for the release of this response. Actually we had not expected much else, for the visual acuity of eyes of the type *Eumenis* has is probably such that, under the conditions of our tests, no details could be visible to the males. We did the tests nevertheless to make sure that we could use the simple paper model as a roughly optimum standard model, with which all other models could be compared.

We next tested the effect of colour, by offering identically shaped models of five different colours—red, yellow, green, blue and brown (the standard colour), as well as pure white and black. These papers were taken from the well known German-made Hering papers, which were standardized, and of which we had measured the spectral reflection graphs. All in all, in 12 test series, we scored over 6,000 reponses and to our amazement there was no appreciable difference between the colours; standard was exactly as effective as red and scarcely different from black, green, blue and yellow. Only white was definitely less effective. If anything there was a trend for the darkest papers to release more responses, but that was all. This we checked in a separate series in which we compared black and white with two intermediate greys. Over 1,000 responses obtained with these papers were divided in such a way that black received most (score 52), grey 15 (the darkest grey) scored 50, grey 7 received 47, and the score of white was 42.

As far as colour and shade were concerned, therefore, it could be concluded that a male was most strongly stimulated by a dark female, of any colour. The colour series even suggested that a black model was slightly better than a model of normal colour.

Now this seemed rather paradoxical. The males did not show a preference for any colour and behaved as if they were colour blind. Yet we had often watched Graylings feeding on flowers and had become convinced that they must be able to distinguish colours, since they showed a preference for yellow and bluish flowers of many shades and hues.

We decided to study the feeding behaviour a little more. To this end we erected a large (5 × 5 × 2 metres) gauze cage in which the butterflies could fly round and visit models of flowers if we wanted them to. We laid out, chessboard fashion, rectangular pieces of paper of all available colours ranging through the spectrum, from red to violet, and all 30 shades of the standardized Hering series. We then captured a dozen Graylings, released them in the cage and, making ourselves at ease and inconspicuous in a corner, awaited events. The sun was hot, the white gauze was incredibly white, the atmosphere very stuffy, and the observers very sleepy. Enthusiasm reached an all-time low when a whole day's session did not yield one single response to any of the papers. At first we thought that the butterflies were not hungry or perhaps disturbed by the unfamiliar surroundings. But when we supplied them with real flowers, most of them did feed.

We then assumed that perhaps, after all, they responded to the flowers' scent rather than to their colours. Their visits to the bleeding trees at any rate were controlled by scent. To test this, we now took some rags, sprinkled them with various perfumes (lavender, rose, lilac) and dropped them in the cage. The Graylings actually responded to those, but in an entirely unexpected way: they walked round, drummed excitedly with their antennae, but they did not go to the rags. Instead, Meeuse suddenly saw one of them fly down to land on the blue shirt he was wearing. Instantly he knew what it meant and kicked himself for not having thought of it before. Actually, a German worker, Dr Dora Ilse, had already reported that the scent of flowers is necessary to some butterflies to make them respond to colours; obviously *Eumenis* behaved similarly. This proved correct and from now on the butterflies were exposed to both scent and papers. The observers, similarly exposed, became almost sick in the heavily perfumed cage, but their reward, after a few days, was an unambiguous graph: almost all of the roughly 100 responses were to two of the papers—blue and yellow. This was a pleasingly clear case of a phenomenon that we know now must be very common: the internal condition of an animal—or its 'mood'—has a profound influence on the stimuli it selects from its surroundings. This particular test showed it more clearly than usual because the papers offered to hungry males were identical in colour and hue to those offered in our fishing rod tests to sexually active males.

At the time, we did not pursue this any further; we just recorded the fact that sexually active males responded to fewer stimuli than their eyes could receive. This phenomenon—response to fewer stimuli than a sense organ can receive—is now known to be quite common. It poses an interesting physiological problem, for it means that somehow information received, and perhaps passed on by the sense organs, can be prevented from ultimately reaching the motor centres of the muscles; and whether or not this happens must be dependent on the internal state of the animal. This problem is now being tackled in various laboratories. Here, as in several other problems, analyses of behaviour done in the field are linking up with neurophysiological research done in the laboratory.

Returning to our fishing rod tests: we next did a series in which models of different shapes were compared—a standard model of butterfly shape, a circular model, and three rectangles of varying proportions (16×1, 8×2, and 5×3.2 cm.). The surface area of all models was the same. The males responded to all these models equally well, except for a slightly lower response to the longest rectangle. However, we found later that the fluttering movements of the models was an important stimulus and as this long rectangle did not flutter as conspicuously as the others, its lower score may have been due to movement rather than shape. In any case it was clear that characters of shape did not contribute to the release of the sexual pursuit. To what extent this was due to inability to distinguish between those shapes at all we do not know.

When we tried a similar series with models of a different size, we ran into difficulties. First of all, when we presented our models in the usual way—making them appear at about 2 yards distance and then approach the male—we found that the effectiveness of the standard model was superior to that of smaller and of larger models. We noticed, however, that the larger models often frightened the males away instead of attracting them. This was particularly noticeable at short distances. We therefore changed our method, models of different sizes being made to appear at a given distance and danced there without approaching. When we did this with a model of about normal size and one of four times the surface area, we got different results depending on the distance: at 50 cm. the large model received about half as many responses as the smaller, and it often scared the male away. At 100 cm. there was no sign of fear and the larger model released more responses than the standard model!

Now a larger model, if presented at the same distance as a smaller one, naturally stimulates a larger area of the retina. Was its greater releasing value due to this or would the difference persist if the two models were seen at the same angle? We did two series to study this—one with a model of 4 cm. diameter at 50 cm. distance and one of 8 cm. at 100 cm.; and another with models of 2 and 4 cm. at 30 and 60 cm. distance, respectively. In both series the larger model elicited considerably more responses than the smaller. This demonstrated at the same time that our butterflies could judge distance or, in other words, had three-dimensional vision.

We did not continue these tests by giving still larger models, so we do not know whether we can describe the result as 'the bigger the better', but the fact that a double-sized model was better than one of normal size was striking enough in itself.

I will not try to describe all the other tests we did, but some of the more striking results must be mentioned. When we showed one and the same model at various distances and made it dance there at the spot, we found that the nearer it was to the males the more responses it elicited. In other tests we examined the influence of the way a model moved. A circular disc was moved overhead either smoothly, or dancing the way a butterfly dances up and down in flight, or turning rapidly round its axis, thus imitating in a crude way the flapping of a butterfly, displaying in quick succession now its broadside, now its narrow edge. Both the dancing and the turning disc elicited twice as many responses as the smoothly sailing model did.

Other tests were designed to study the interaction of the various effective stimuli—darkness, nearness, largeness and movement—and the general outcome was that a deficit in one stimulus (e.g., darkness) could be compensated for by an increase in another one (e.g., nearness). This showed the curiously automatic nature of such responses. Looking at it anthropomorphically, one would expect that an abnormally coloured model would release fewer responses the nearer it was offered. Yet exactly the opposite happened, the 'worse' a model, the nearer it had to come to release responses.

These observations raise a number of questions and suggest a great variety of further tests. For instance, is the greater effectiveness of the black models due to their contrasting with the sky? Would white show up best when all models were presented against a black background? What is the effect of 'turning' and 'dancing'? Is it just the speed of partial movements in both cases and the succession, on each part of the retina, of dark and light? These questions have not been answered yet; in fact, we are only just beginning to see which are the general problems raised by this work. One is: how are these stimuli, as we discover them in these tests, organized? It is obvious that each of these so-called stimuli set off complex, synthetic processes, in which separate sensory data are integrated in data of a higher order ('dancing'; size-independent-of-subtended-angle).

Another question has already been mentioned: how are some of the sensory data prevented from influencing certain responses? It is quite possible that we have reached the limit of field work here and that these problems have to be tackled in the lab. In fact that is where they are being tackled now.

The sexual pursuit is only one link in the total mating behaviour of the Grayling. Analyses of the other links in the chain were started by presenting models of females, mounted on wire, to males that had first been lured to the ground by a flying model. Unfortunately this work, although it was carried on for ten seasons, was never completed. Yet a number of interesting things were discovered. For instance, while the male is performing the whole sequence of his courtship, he faces a stationary and motionless female all the time, yet each link of his action chain was a response to different stimuli provided by the female. Some links were evoked by visual stimuli, others by the female's scent as well. Further, the female's antennae were necessary for the male's orientation when he positioned himself in front of her. We could show this by offering a dead female of which the antennae had been removed, and two pins stuck into her abdomen imitating antennae at the wrong end of the body. The male then took up a position behind the model instead of in front. Yet when he walked round to copulate he did not orient himself to the antennae and with a model with 'antennae' at the wrong end he copulated in the normal position.

The most spectacular part of the male's courtship was the 'bow' and we wondered what its function might be. The behaviour of the female gave us the clue. When the male makes his bow, the female's antennae are stretched towards him and their tips caught between his fore wings. Sometimes a female tolerated the male's antics for a while, but flew away before he could copulate. This refusal occurred often exactly at the moment when the male's wings touched the clubs of her antennae. It is in these clubs that the scent receptors are situated. Watching the male's bow carefully, we found that the female's antennae came in touch with the male's wings at the spot where each male had a dark patch which was not present in females. Under the microscope we saw that this patch was a field of aberrantly shaped scales; they were not flat and broad, but slender and ended in a tiny brush. This type of scale is well known and has been found in many species of Lepidoptera, always being associated with the production of scent. Often this scent is quite distinct, even for the insensitive human nose and in some species it is quite pleasant. The great naturalist, Fritz Müller—who was the first to describe these scent scales—mentions how, on his journeys in Brazil, he often carried a *Papilio Grayi* with him, just for the purpose of sniffing its scent when the mood took him.

The Grayling has been reported to give off a distinct smell of chocolate, but we could never be sure that we could smell anything at all. Yet as the indications were strong that the male's bow had the function of presenting a chemical stimulus to the female, we decided to test this. We found soon that this was more easily said than done and the cost and energy spent on the test were perhaps out of all proportion to the meagre result.

We started by catching a fair number of males and females as early in the season as possible. Half the males were subjected to a harmless little operation: with a fine brush their scent scales were removed and the area they had covered was then painted over with an alcoholic shellac solution. The other males received a similar treatment but involving an area just beside the scent field, which itself was left intact. All these males were then released in the cage and half as many females allowed to join them. Hoping that they would mate under these conditions we then watched whether the females would show any preference. The weather, which was usually unstable, and the difficulty of finding sufficient virgin females, made progress very slow, but gradually we collected evidence on 27 matings observed in the cage. Of these, 16 were with males whose scent organs were intact, 11 with de-scented males. This was a poor and not very clear-cut result. However, we had seen that several of the 11 matings of the scentless males had occurred in an irregular fashion: a female was being courted by a scented male facing her, when suddenly an unscented male walked up and copulated with her. This happened in the corners of the cage, where the Graylings often crowded together. It looked, therefore, as if scentless males were sometimes accepted by females that had been aroused by normal males.

We now did our next tests in two cages. The scentless males were in one, the normal males in another. This gave us a score of 9 to 2. Altogether, therefore, we scored 25 versus 13 and since we knew that at least 5 of these 13, and probably more, were 'stolen' matings, there can be little doubt that the scent scales facilitated mating by making the females accept the males. The lack of success of the scentless males was not due to a possible detrimental influence of the operation on their courtship behaviour; if anything they courted more vigorously than the normal males, presumably because they were continually thwarted by the females' refusals.

This result, meagre though it was, pleased us because it suggested another type of function of scent scales than that usually assumed. In many species scent serves to attract the other sex; here the sexes came together by the male responding to the sight of a passing female and the scent did not come into play until the final stage of courtship; its function was to release the female's co-operation in actual mating.

Thus the Grayling had not disappointed us, but gave us plenty of material for study. Much of our work, interrupted by the war, was not taken up again; and so we did not continue our analysis of the males' behaviour; and our microscopical observations on the scent glands and on the olfactory organs on the antennae, and a morphological study of the copulatory organs of males and females of the Grayling and related species were never finished. The incompleteness of the results is not due to the Grayling but to us; and a return to the Grayling is one of those *desiderata* which one refuses to write off.

CHAPTER TEN

CAMBERWELL BEAUTIES

One sunny afternoon in the first half of July, while walking over the sands in search of male Graylings—needed for our fishing rod tests—we came upon a large spiny caterpillar, black with a row of dull orange dots, which was crawling over the bare sand at what was, for a caterpillar, a terrific pace. It covered about a metre a minute. We recognized it as the last instar of the Camberwell Beauty, a lovely *Vanessa* which is not rare in the Hulshorst area. The nearest Birch tree (the food plant of this species) was 60 metres away and the larva was running away from it. We put it on the tree, but it refused to stay. Obviously it was due to pupate.

We knew that *Vanessa antiopa* caterpillars, like those of their relatives the Small Tortoiseshell and the Peacock, live in clusters. After a little searching in the neighbourhood we found the whole family in a small Birch—62 large caterpillars.

When we touched this tree, a couple of larvae dropped on the ground and each began at once to crawl away; they seemed just as much in a hurry as the one we had seen first. Another slight shake and five more dropped off. They too started to travel. None tried to get back to the tree.

We decided to follow some of them. Their pace, about 1 metre per minute at first, soon slackened down to about half this speed and after they had covered about 80 metres (which took them slightly over an hour and a half) they crawled at less than one tenth of their original speed. Their course had been practically straight, but towards the end each larva changed course and made for the nearest tree or bush. Slowly each of them crawled up the tree of its choice and came to a rest on one of the lower branches. Some of these larvae had passed several trees on their way, but had ignored them. Clearly their interest had changed towards the end of their journey.

Although each caterpillar followed an almost straight course, those of the different larvae were different—they radiated out from their food tree in various directions. Nor did they demonstrate any sign of preference for certain trees; it seemed as if any tree would do once the larva was in need of one.

By the end of this afternoon all the larvae whose start we had witnessed had arrived at one tree or another and all were now clinging motionless to their branches.

Next day the weather was cold and rainy and, our *Philanthus*- and *Satyrus*-crews being unemployed, we decided to return to our *antiopae* and watch them a little more. When we arrived, about 9 a.m., most of the caterpillars had left the Birch and were crawling over the sand. Because of the cold they were very sluggish, but we decided to follow them. To our surprise their tracks were entirely different from those of yesterday's travellers; they wound round and round in quite irregular fashion, often crossing their own tracks more than once. Some, having wandered over the sand for an hour or so, arrived back at their starting place and went on in another direction. It was a scene of hopeless confusion. It continued for about four hours.

In the afternoon the cloud cover which had been complete, broke. Now and then a light area appeared and later the sun came through. As soon as the light area in

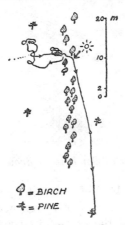

Fig. 18. Track of a caterpillar of the Camberwell Beauty; the arrow indicates the moment the sun came out. Migration started from black Birch.

the clouds appeared, each caterpillar began to follow a straight course, each in a different direction. When the sun warmed the ground, their pace began to increase as well and soon they were all behaving exactly like their brothers the day before: all ultimately steered towards a tree and climbed it.

It seemed obvious that the caterpillars took their bearings from the sun, although each in a different way, but when we did a crude test to check this, shading a larva while reflecting the sun on them by a mirror held opposite the sun, we could not make them turn round. We had at that moment no means of doing anything more precise about this. It is possible that our hand, used to screen the sun, was too small and that the bright area round the sun could still compete with the sun's reflection, or the larvae may have responded to the polarization of the blue sky—a possibility which at that time did not occur to us.

Fig. 19.   Caterpillar of Camberwell Beauty ready to pupate.

We did another kind of test with more success. When we put up a tree or part of a tree near a larva which had about reached the end of its travels, it invariably turned off towards the tree and started to climb it. It was even quite sufficient to post ourselves near such a larva; it would respond to us as if we were a tree and climb us. We could steer them in any direction we wanted. This little ruse, however, had not the slightest effect on larvae that had just started; their interest in trees came at the end of the migration.

When the day drew to a close, 45 caterpillars had reached a tree each. For about a day they remained clinging to the underside of a more or less horizontal branch. Here they spun a little mat of silk and, when this was ready, hung themselves up by hooking their posterior legs in this silk mat. In this position, they bent their thoraxes, assuming the shape of a J. These J's remained hanging rigidly for 1–3 days. Then they pupated. We observed this several times and, although it is quite a well-known process, yet every time it is fascinating to watch. When about to pupate, the larva makes a few convulsive movements, after which a series of contraction waves run forward over the whole body. The abdomen becomes thinner and thinner, the thorax swells. Suddenly the skin on the back of the third ring cracks.

And then you suddenly realize that you are not watching the pupation at all, but merely the very last phase: while hanging motionlessly, the larva has grown a pupa's skin under its black coat and now it just shakes off the old coat. Through the crack the new skin, pale buff in colour, protrudes. Very slowly the black skin is now moved back and peeled off. More and more of the pupa shows; very slowly it wriggles out of the old skin.

But this leads to a critical situation: the larva is hooked up by its outer skin and unless the pupa can attach itself to the silk mat, it is bound to fall on the ground. This tricky manoeuvre is usually done very cleverly: with the sharp spiny protrusions on the abdomen, particularly with the last pair, the pupa remains hooked in the inside of the larval skin, while the tip of the pupa's abdomen protrudes into the air, making searching movements until it touches the mat. Then, by a series of jerky semi-circular motions, it entangles its tip—which is covered with a field of very fine hooks—into the silk. It then discards the old coat, which at first remains attached to it like a black collar, but soon drops off.

A few days after we had observed the pupation journeys all caterpillars had pupated. The pupae, a pale brown or buff at first, soon turned a darker brownish khaki and were then exceedingly difficult to find. As is known of other *Vanessa* species, the individual variation of colour is large; with our pupae there was a clear correlation with the colour of the background, the over-all result being that most of the pupae were beautifully camouflaged—quite different from the larvae, which were always visible from afar. The 45 pupae were scattered over a number of trees in an area of approximately 6 acres.

We checked the pupae from time to time and after 20 days the first hatched. Two days later all of them had emerged, except four, which had been killed by a parasitic fly, probably *Sturmia pupiphaga*. Three of these had been hanging on the same tree and no pupae on this tree had escaped.

We never followed up these incidental and, I admit, slight observations, but they made us aware of several interesting points. One concerns the orientation of the larvae during their pupation journey. It was obvious that an overcast sky disorients them completely. The sun, or even just a bright patch in the clouds, allows them to follow a straight course. The mirror test suggests that the sight of the sun itself is not absolutely necessary—perhaps they can use the polarized sky.

Another striking point is the curious contrast between the sociability of the caterpillars and the sudden individualism of those about to pupate. Why should each larva embark upon this long and risky journey? I believe that we have to do with a change from one type of 'defence by colour' into another and that the changes in behaviour have to do with these changes in colour. The black caterpillars, which are not camouflaged, live in clusters. Although no experiments have been done with our particular species, it has been shown that the closely related and very similar larvae of the Small Tortoiseshell and the Peacock do derive protection from living in clusters. Redstarts for instance do not as a rule attack these larvae while they are clustering, but as soon as one detaches itself from the group and wanders off, the birds are far less reluctant to take them. Why the birds behave in this way, we do not know—but the fact is there. It seems likely that the crowding of the *antiopa* caterpillars has the same function.

The pupae, however, are not conspicuous and black, but camouflaged. It seems to me that this must be correlated to the need for avoiding the crowd. Most camouflaged animals live scattered and, as I mentioned in Chapter Seven, there is evidence showing why this is

useful: some predators, at least, are encouraged to concentrate their attention on one kind of prey if they find a
few individuals of this particular species in quick succession. They give up, however, when after one lucky
find, their further search does not pay. Even a predator,
or for that matter a parasite, which specializes on one
prey would probably be more of a menace to crowded
than to scattered animals since it may tend to keep
searching near a place where it has had success. The fact
that 3 of the 4 parasitized pupae were in one tree might
be due to something of this kind.

Scattering, therefore, looks as if it is an adaptation
which prevents predators from being too persistent and
from specializing. The fact that these vulnerable larvae
take the risk of this pupation journey in order to get away
from their brothers and sisters gives us an idea of the
extreme importance of the effect—dispersion. The Camberwell Beauty and its relatives are particularly striking
by their sudden and dramatic transition from one type
of adaptive coloration to another, together with the correlated behaviour. Our observations not only impressed
on us to what lengths a species may go just for the purpose of protection from predators, but also the curious
and altogether puzzling fact of the wholesale change
from one system (conspicuous colour; crowding) into another (camouflage; spacing-out), with all the trouble it
entails.

CHAPTER ELEVEN

CLIFF DWELLERS

To acquire an understanding of how birds live, and why they behave as they do, you can select any bird you happen to see. House Sparrows are as interesting as Golden Eagles. But everyone has his preferences, often depending on opportunity, accident and various imponderables. I happen to like sea birds; first of all, I think, because as a boy I fell under the spell of coastal scenery; further because many of them are large, live in open country and are, therefore, easily observed and are social. Song birds do not appeal so much to me. I hate craning my neck all the time to follow them moving through the tree tops and to miss the most interesting things because they happen behind some thick branch. Incidentally, not all sea birds allow one to do so without craning one's neck. I have vivid and rather painful memories of the seasons I spent watching the aerial displays of terns.

Among the sea birds, gulls are my favourites. While living in Holland I had, in the 'thirties and 'forties, spent much time on a study of the behaviour of Herring Gulls, more particularly of their social organization. The results, though slow to come, had been so fascinating to me that I began to wonder about other gulls. Also my friend, Konrad Lorenz, had often urged me to start comparative studies of a group of closely related species, reminding me of the splendid results of comparative anatomy, which had contributed so much to our insight in the evolution of living beings—in their true affinities and in the way related species must have diverged gradually from common ancestors. Lorenz often pointed out that we still know next to nothing about the evolution of behaviour and that comparative behaviour studies are urgently needed, if only to provide the necessary descriptive basis for a study of the dynamics of behaviour evolution.

When, in 1949, I settled in Oxford and had to find opportunities for field work for graduates, I naturally thought of gulls. But as Sir Alister Hardy, the head of the Zoology Department at Oxford, pointed out to me at the time, Oxford is about the worst possible place in the British Isles to satisfy such an ambition: coasts (where most gulls live) were not exactly in our back yard, as the North Sea coast had been at Leiden. However, when he saw that I meant to go ahead with gulls, he supported me fully. The Department of Zoology acquired a Land Rover for transport, which radically changed the position. As Professor Hardy likes to say now, all the gulleries in Britain are equally near to Oxford and our place is really the ideal centre for a 'gull programme'!

It was not difficult to attract collaborators and soon four were ready to start—Martin Moynihan of Princeton University, U.S.A., Dr Esther Cullen of Bâle, Switzerland, Dr Uli Weidmann of Zürich, Switzerland, and Mrs Rita Weidmann of Oxford University.

Our first task was to decide which species we would tackle, and where. Not all gulleries were equally suitable. We had to have large, easily accessible colonies, with camping or living quarters nearby, and they had to be protected so as to be free from interference. Also, the owners and those responsible for the protection would have to allow us to put up our hides and to carry out our observations and certain experiments, which naturally would cause some disturbance. In other words we re-

quired permission to do exactly those things which protection meant to prevent. We were extremely fortunate
in getting access to ideally situated colonies of Black-
headed Gulls and Kittiwakes—both very different from
the Herring Gulls and yet related to them.

Just when we were considering ways and means to get
going on this programme, we met Dr Eric Ennion, the
well-known director of Monks' House bird observatory in
Northumberland. 'If you are so interested in sea birds,
why don't you come to the Farne Islands?' he said. 'We
live on the coast just opposite them and they are just
covered with birds—literally!' And he began to rattle off:
'Shags, four species of terns, Eider Ducks, Fulmars,
Puffins, Razorbills, Guillemots, Oyster-catchers, Herring
Gulls, Lesser Blackbacks, Kittiwakes!' And when he had
got his breath again he added: 'And you can leave your
hides at home, because they are all as tame as chickens.'

This sounded fantastic, but we soon saw for ourselves
that it was true.

On a brilliant day in June, 1952, we paid our first visit
to the Farne Islands. In one of the 'cobles' that the local
fishermen use for their crab- and lobster-fishing, we set
out from the small tidal harbour of Seahouses. Scarcely
half a mile out, when we were approaching the group
of tiny off-shore islands, we began to see the first sea
birds. Stately Gannets, in their spotless white plumage
with black wing tips, passed us in majestic flight, occasionally plunging down into the sea to levy their toll
of the numerous small fish. Shags and Cormorants flew
hurriedly by, some of them carrying huge bundles of seaweed to their nests on the distant cliffs. Kittiwakes and
various terns delighted us by their vigorous and elegant
wing beat. Grey Seals stretched their necks out of the
water to have a good look at our boat before disappear-
ing again under the surface. And near the islands the sea

was absolutely covered with Puffins, Guillemots, Razor-bills and Eiders.

We made what we learned later was the usual trip—for many people come to have a look at the islands. First we went to a low skerry, the Crumbstone, where about a hundred seals, a fraction of the total population living round the islands, were basking on the rocks. Interesting though such a crowd was, they were very much disturbed by us and all we saw was what looked like a horde of puddings waddling to the water and splashing in, and after that a number of shiny, bobbing heads, astonishingly dog-like, looking curiously at us. Later, we got to know some of them quite well, particularly one huge bull who used to catch his breakfast, in the form of a couple of large Lumpsuckers, right at the foot of the cliff where we did our observations; we named him after one of our colleagues to whom he bore a striking resemblance.

From the Crumbstone we went to the Staples and, after cruising for a while under the magnificent Pinnacles, three basalt stacks topped with hordes of Guillemots, we landed on the island, and had our first view of what could be called a glorified zoo—Kittiwake Gully. On the steep walls bordering a very narrow cleft row upon row of Kittiwakes were sitting quietly on their nests like tins on a grocer's shelves. We walked up and sat down within a few feet of them, but they never cared and just went on with their business, which was feeding their lovely silvery grey chicks, dozing, or fighting each other off. Between them there were Shags, slender, dark Cormorants with a fantastic green sheen on their necks and bronze-coloured, scale-like feathers on their backs. When we came near, they would not budge, but threatened us with a curious darting movement of the head, with their bright yellow mouths wide open. While sitting there in the brilliant sunshine and drinking in the beauty of the scene, we heard, just behind us on the Pinnacles, the

waves of curious mutterings that signified that the Guillemots were having arguments among themselves.

After a few hours on the Staples we went to the
Inner Farne. There we were welcomed by a cloud of
screaming Arctic Terns, which occupied the tiny sandy
beach in St Cuthbert's Cove and the adjoining rocky
shore. Eider Ducks with their charming downy chicks
were everywhere, with the ferocious Herring Gulls and
Lesser Black-backed Gulls in attendance, ever keen for
a tasty duckling. After having a look at the squat medie-
val tower, the old chapel and some ruins of the time
when the island was inhabited by the Holy Island monks
(successors to the famous St Cuthbert), we walked up
the gentle slope to the south-westerly top of the island,
where we had our great surprise. From a height of about
60 ft. the land fell off abruptly, forming a steep cliff,
which was occupied by some small colonies of Kittiwakes, with an occasional pair of Shags, some Razorbills
and Guillemots, and a few Fulmars. On both sides of the
cliff several hundred Puffins stood eyeing us distrustfully.
At the foot of the cliff the bluish green sea washed the
barnacle-covered rocks and here and there great forests
of golden Laminaria were lazily rocked on the gentle
waves.

This was the island where, we were told, the Farne Islands Committee were planning to establish a Field
Study Centre. Visiting students would be provided with
living quarters in the top floor of the old tower so that
they could live on the island for weeks on end. Then
and there we made up our minds that we would try to
be the first visiting students. Esther Cullen fell in love
with the Kittiwakes at once—which suited me very well,
for we hoped that the Kittiwake, a rather aberrant species of gull, would show interesting differences from the
large gulls we knew. As we will see later, the results of
her work were even more striking than we could have

expected. Mike Cullen saw his chance to study the Arctic Tern here, in a colony of well over a thousand pairs.

The Farne Islands Committee (Sir John Craster, Dr Eric Ennion and Mrs Grace Hickling—Miss Grace Watt, the author of *The Farne Islands*) kindly allowed us to spend three weeks on the island that summer to make some preliminary observations and even put the tower at our disposal for three full seasons afterwards. Thus we joined that curious little guild of British island-dwelling naturalists who have all spent some of the best years of their lives on similar isolated spots. Sometimes we meet other members of the guild—R. M. Lockley, Frank Fraser Darling, or Kenneth Williamson—and, although life on the Inner Farne cannot be compared with that on Rona, or even Skokholm or Fair Isle, we perfectly understand each other when we are thrown together in a stuffy town: we long to get back to our islands, to feel the sea breeze, to see the waves breaking over the rocks, to hear the cooing of the Eiders and the Kittiwake-ing of the gulls echoing against the cliffs, to sniff the invigorating salty air, loaded with the sweet scent of Sea Campion or with the typical, though far from sweet, smell of the high tide mark.

The Cullens spent in all three seasons there, usually from early March until August. I myself accompanied them for periods of a fortnight at a time, several times a season, and busied myself with cooperating in a small way in their work and with making photographic and film records of their studies.

When we arrived at the end of the winter we usually spent one night in hospitable Monks' House, laid in food stores for three weeks, and then got in touch with the Shiel family—the late Scott, his brother Jack, and Jack's son Bill—who, weather permitting, loaded our bulky gear into their boat and ferried us over. The Shiels also, by arrangement, supplied us with fuel, drinking water and occasional sea-food, and took over our weekly rations and our mail. Their visits were extra welcome after a prolonged gale (the island was occasionally inaccessible for periods up to a week) and when we had run out of tobacco, for instance, the name of their coble, 'Glad Tidings', was very appropriate.

Life on the islands was delightfully uncomplicated. We often got up just when the very first glow of light appeared on the eastern sky, had a cup of coffee and a slice of bread and then set out for the cliff. Mike Cullen, whose terns did not arrive until the beginning of May, spent the first two months of each season watching Shags.

Observing the birds on the cliff was cold work. Although it was rarely freezing, even temperatures of 40 degrees can make one quite miserable when they go with clouded, windy and wet weather. The first hour would not be bad, but then the cold began to creep in slowly but steadily; and there was nothing you could do against it, for bird watching of this kind just means sitting still. We were armed with heavy clothing: layer upon layer of wool, with duffelcoats on top, covered again by windproof coats. In order to keep our hands warm enough for fast writing, we had heavy fur-lined gloves, yet at the end of four hours' watching none of our joints could move as it ought to. Esther found the best solution at the end: wrapped in all those clothes, she wriggled into a sleeping bag, balancing precariously on the edge of the cliff, and then managed to squeeze in a hot-water bottle as well. We were all for comfort!

When the early hours were over, we had a break, birds permitting. On our way back to the tower, we kept our eyes open for whatever migrant song birds might be about. Once upstairs, it was the work of a few minutes to get the Primus going and then we had a leisurely breakfast, looking out of the window and watching the wide sea and the strip of coast in front of us. A book could be written about the things seen from that window alone: seals came up with their catch, harassed by screaming gulls; Gannets fished right under the shore, Oyster-catchers fought their battles on the rocks, Ringed Plovers nested not far from the window, and the Eider Ducks settled by dozens in our yard and sat patiently for a month, after which they suddenly produced from under their broad bellies a clutch of lovely downy duck-lings. But I will tell the full Eider story in Chapter Fourteen.

After breakfast, when various household chores had been done, the rest of the day was again spent on the cliff or elsewhere on the island. A folding canoe allowed us to pay occasional visits to other islands.

The tower had its disadvantages. On the floor under our apartments the bird wardens lived from the begin-ning of May on. We were separated by just one layer of not too well fitting planks and our neighbours could not help hearing our every whisper. Worse, all the water spilled on our floor found its way down through the cracks in no time, cries of distress reaching our ears the moment a bucket was kicked over. Judging from this, our neighbours seemed to live in constant dread of us. On the whole, however, we behaved ourselves and our neighbours were always remarkably tolerant. This turned into outspoken gratitude after we had been joined for a week by some colleagues from another university, who, though intelligent people, did not seem to grasp the causal relation between walking round on nailed boots at 4 a.m. and the suffering countenances of our friends the wardens later in the day.

Another slight snag was the presence of many cracks in the 3 ft. thick walls. One of them was so big that one day a Robin that had blundered into the room found its way out through it--at least it disappeared in a dark corner and was never seen again. Through that same crack we heard the Starlings outside sing as if they were sitting right in the room. In May their clamouring young made themselves heard with daily increasing emphasis, until they forced us to stop talking whenever they thought their parents were coming in with food. But there were numerous smaller cracks as well. Although through none of them could you see the outside world, the northerly gales found their way in. I still vividly remember Coronation week, which found us shivering in front of a blazing fire, with all our outdoors clothes on, the heat all being blown out through the chimney by a terrific cold blast entering through the cracks and passing us on its way to the chimney. It was the traditional English open fire carried *ad absurdum*—delightful! We often practised enjoying it, standing with our back to the fire. Here I learned that one has to start being British quite young!

When we started our observations, in the first days of March, there were practically no birds on the island and, apart from an occasional Fulmar or a Guillemot, the cliff was bare. But the Kittiwakes were already in the neighbourhood and were showing an interest in the cliffs. Often a flock of ten or twenty would settle on the water in front of the cliff, staying a couple of hundreds of yards out; obviously the island began to attract them, but for some reason they were not yet prepared to settle on the cliff. While they floated thus in front of us, they continually called their quarrelsome 'kittiwaak'-call. When the tide was running—which it did most of the time—it soon drifted the flock away from the nesting cliffs on the island. The birds were carried further and further away, until they were mere specks in the distance. Suddenly, one bird would fly up, then another, and another; and soon the whole flock would come flying back and settle once again in front of the cliff. Not until several days later did the first birds alight on the cliff itself. Often such a pioneer was one of the dozens of birds around. It would never stay long and, while it was on the cliff, it showed every sign of being ill at ease, with sleeked plumage and stretched neck. But even though it would soon leave again, it would return; and soon more birds came and occupied the ledges. They selected the steepest parts of the cliff, preferring the absolutely sheer walls, where they settled on very narrow 'sills', often no more than four inches wide.

This was when observations started in earnest. The first thing to do was to get to know individual birds, so that they could be recognized even after prolonged absence.

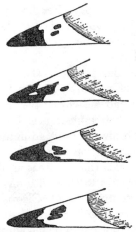

Fig. 20. 'Identity cards' (wing tip patterns) of four Kittiwakes.

With other birds, one would do this by capturing them and marking them, for instance with coloured rings on the legs. But catching adult Kittiwakes on any scale was not so easy. We might have tried it but at best would have had very meagre results even after spending considerable time and energy on it.

Mrs Cullen found a much better way: she noticed that the pattern of the black dots on the wing tips differed considerably from bird to bird and, after some practice, she could recognize a number of them by these 'identity cards'. Soon she had found a good lookout post opposite a cliff inhabited by about 30 pairs of Kittiwakes. Because she was sitting there quietly day after day the birds soon got used to her and finally ignored her altogether. Her only equipment, apart from field glasses and note-book, was a chart with drawings of the wing-tip patterns of those birds she had become personally acquainted with. She found that the patterns remained distinctive even through the moult and so she could recognize her old acquaintances every spring, after being separated from them through a whole winter. Most of the birds returned year after year to the same part of the colony, usually even to the same ledge.

All through the four seasons (she spent three entire seasons and part of a fourth summer) she found that many of the birds had individual peculiarities in behaviour as well as in their appearance. For instance, there was one pair which always built an unusually high nest; another bird, a female, was too shy to mate; although she kept visiting males through season after season, she was always too nervous to stay with any of them. (This bird was inadvertently named Cleopatra before her character was known.) Other birds had peculiarities in their calls by which they could be recognized.

On the whole, it was found that the social organization of a Kittiwake colony was similar to that of other gulls. The birds were strictly monogamous and pairs kept together throughout the season, often even through more than one year. Mates recognized each other personally. They formed pairs on the ledges, built nests together; the male fed the female; the partners took turns in incubation and in feeding the young. There was also evidence showing that neighbours often recognized each other. The various postures and calls acting as signals or 'language' were roughly the same as those of other species of gulls.

However, there were also many points in which the Kittiwakes differed from other gulls and on these Mrs Cullen concentrated. Because her observations throw light on the nature of these differences and have a bearing on problems of evolutionary divergence, I will discuss this aspect in some detail.

First of all, Kittiwakes are remarkably tame. That is true not only of those on the Farne Islands, but of Kittiwakes of other localities as well. They not only tolerate people near their nests, but care amazingly little for other predators either. The difference, for instance, in the way they respond to those notorious robbers of eggs, the Herring Gulls, and the way Black-headed Gulls or Common Gulls do is striking: whereas those other species utter a high-intensity alarm call whenever they see a large gull in the distance and fly up and attack it furiously, we never saw Kittiwakes pay any attention whatsoever to the Herring Gulls. They are so tame that it is even difficult to hear their alarm-call at all. We did not hear it for weeks, until we decided to climb down the cliffs and actually visited the nests. Mrs Cullen realized that this tameness had to do with the habit of breeding on steep cliffs. There is no doubt that this cliff-breeding habit is a method of protection against predators. Few mammals ever venture on such cliffs. We have no evidence, it is true, that a Fox avoids cliffs, because we never saw a Fox near Kittiwake colonies, but in the Arctic the small but fierce and tireless Arctic Foxes roam all over the country in search of eggs and young birds, and they avoid steep cliffs. Even the Herring Gulls seem to have difficulty in alighting on such cliffs, and as far as we know our colony did not lose a single egg to Herring Gulls, although the latter were around all the time and fed regularly on more readily available eggs such as those of the Eider Ducks. A species such as the Kittiwake, which has such an excellent defence against predators, has no need for escape and alarm calls and social attack; it need not waste time on flight or attack and can spend the time other species have to give to these emergency responses on other activities instead. They are like a man living in a fire-proof house, who need not spend time earning money to pay his fire insurance premium, so can use extra money for some other purpose.

Another peculiarity of the Kittiwakes is their excessively quarrelsome nature. At least in the breeding season they spend an enormous amount of time fighting and threatening each other. Other gulls do fight a great deal, but not nearly as much as Kittiwakes. The functional significance of this becomes clear when you compare the types of breeding grounds of other gulls with those selected by the Kittiwakes: their choosiness in requiring not only steep cliffs but narrow ledges as well (which are usually restricted in number) makes them the only species of gull which faces a housing shortage. The other species we have studied so far nest on the ground and usually have an unlimited number of adequate nest sites available and gulls that are driven off already occupied ground simply move on a little and find another suitable place.

Kittiwakes are not only restricted by the type of the habitat they prefer, but they also want to nest near others. Very few of them will ever choose a suitable ledge far from the main colony; rather they will fight for weeks and weeks against overwhelming odds to try to conquer a strongly defended ledge in the middle of the colony. Fights are still frequent towards the end of the breeding season and at least on the Farnes a good many

Kittiwakes during the course of a season never acquire a ledge of their own.

The hostile postures which Kittiwakes assume when they face rivals on the cliff are on the whole similar to those of Herring Gulls and the other species we have studied. Yet there are some interesting differences. One is that the 'Upright Threat Posture', which can be seen in other species, is absent in the Kittiwake. It seems as if this, again, has to do with the Kittiwake's cliff-breeding habit. In my book on the Herring Gull I have tried to show that the Upright Posture is nothing but a combination of parts of movements belonging to two different actions—attack and escape. The stretched neck and the downward-pointed bill can be recognized as the beginning of pecking down at the rival and the posture often continues into an attack of this type. In ground nesting gulls it is usual to attack an opponent from above, hence the neck is stretched upward preparatory to attack.

In the Kittiwake, however, a neighbour can be above one just as well as below one, consequently a standardized method of attacking from above would not be likely to be successful. As a matter of fact, a Kittiwake attacks just as readily upcliff as downcliff. If the interpretation of the origin of the upright posture in other gulls is correct, one would not expect this posture to have developed in the Kittiwake at all, since the full attack, of which it is a part, does not occur in this way. The absence of this posture in the Kittiwake, therefore, is consistent with the interpretation of the origin of the posture in other gulls and with the idea that many traits of the Kittiwake have to do with its cliff-breeding habit.

I will refrain from a discussion of the various other postures shown by the Kittiwake either during hostile encounters or during pair formation, however interesting they are, but will continue our examination of the specific Kittiwake-traits.

Pair formation takes place on the ledges. As in the Black-headed Gull (which I will discuss in some detail in the next two chapters) the formation of a pair is a slow process, in which mutual distrust must gradually give way to toleration and perhaps affection. The first sign of this mutual adjustment is that the male pecks his wife less often than at first, and soon the female no longer stretches her neck nor assumes an anxious posture in the presence of her husband. On the contrary, she withdraws her head until it is resting on her shoulders and, in this 'hunched posture', begins to perform curious head-tossing movements, giving a little squeak every time she throws her bill up. This has an immediate effect on the male: his neck begins to swell and soon he regurgitates food, which the female greedily devours. This 'courtship feeding' occurs in all species of gulls, but again the Kittiwakes' method is slightly aberrant. The Kittiwake male never drops the food, but keeps it in his throat, the female pecking right into his throat as soon as he opens his mouth. Even when she fails to do it, he does not drop the food on the ground, as all other gulls do, but swallows it again. This difference in the technique of feeding, slight though it may seem to be, is absolutely constant, and it seems to have adaptive significance. As other gulls have large territories, it can do little harm when food is dropped on the ground and some remains are left—whatever is lost in this way is scattered over the whole territory. But the Kittiwakes spend all spring and summer on their tiny ledges and wasted food would collect and rot there if always left at the same place. In fact, the whole territory is nothing more than just the nest site and nest sanitation, or rather territory-sanitation, is much more urgent here. Later, the young are fed with just the same precautions.

After the pair has been formed, the birds often wait for several weeks before taking the next step. They spend part of their time on the ledge, part away from the colony, presumably fishing at sea. But one day they suddenly start building a nest. This again is done in quite a peculiar way, different from the ways of the other gulls. Most, perhaps all, other gulls start by making a 'scrape', a rounded shallow pit which they rake out by sitting down at the nest site, shovelling soil, dirt and plants away backwards. The Kittiwakes, nesting on the narrow, rocky ledges, start by collecting wet mud or seaweed; stamping this down by continuous trampling movements of the feet, they work it into a firm platform of hard and strong material, which sticks to the rock. This platform serves two ends—it often broadens the foundation for the nest and it has a horizontal upper surface and so improves the foundations of nests built on slanting ledges. Again, this peculiarity of the Kittiwakes' behaviour fits their peculiar nesting habitat remarkably well.

Mrs Cullen found other differences when she observed the behaviour of the chicks. The young of other gulls begin to leave the nest and walk round shortly after hatching, after one or two days; within a week they wander over quite an area. Kittiwake chicks stay put. They just lie or, later, stand on the nest, often facing the rock, and never walk away. Occasionally the Cullens took a young bird into the tower, and they never had a less peripatetic guest: if put anywhere on a table it just stayed where it was. This is again useful, in fact a necessity, in a cliff dweller.

It was naturally interesting to know whether this difference between young Kittiwakes and other gulls was innate or perhaps imposed upon them by the different environment in which they found themselves. To find this out, the Cullens put some eggs of Black-headed Gulls and Herring Gulls in Kittiwakes' nests. They were accepted and duly hatched, but most of those chicks did not survive very long, for, showing not the slightest inhibitions, they soon began to walk about, with disastrous consequences. The converse test, having Kittiwakes' eggs hatched in the nest of a ground-breeding gull, was not possible on the Inner Farne, but reports have been published about a small Kittiwake colony on a Danish island, where, for some unknown reason, Kittiwakes nest on flat ground. Salomonsen, who has visited this colony, comments on the fact that even here the chicks stay in the nests, under circumstances where chicks of other gulls would move about. There is, therefore, no doubt that we have to do with a real, innate difference between Kittiwakes and other gulls.

Kittiwake chicks are unique among young gulls in other respects, too. For instance, they are not camouflaged; instead of the buff ground colour and a pattern of irregular dark dots, usual in the downy plumage of gulls, they have a beautiful silvery sheen. The next plumage is also quite unlike that of other gulls; it is pied and very conspicuous. Again, in a species whose broods are not subjected to predator pressure, there is no need of camouflage.

Like other young gulls, the young of the Kittiwake try to rob their nest mates of food. This again raises a problem. Other species solve it in a simple way: the young that has been so fortunate to get the bulk of the food often runs away from the others. This would not do for the Kittiwake. Instead, the young that is being attacked for whatever reason turns its face away and this gesture has a remarkable effect: it stops the other from attacking. The same gesture is employed by the adults as well —I will discuss this 'appeasement posture' more in detail in the next chapter. It may be significant that the Kittiwake is the only species in which the young bird has a conspicuous black band across the neck; this is exposed and turned towards the attacker when the face is turned away and, in view of what is known about the signal function of similarly conspicuous colours in so many other animals, it is possible, and even probable, that the black band is just such a signal structure.

Thus Mrs Cullen's study of the Kittiwake revealed a number of properites peculiar to this one aberrant species. On the one hand, it is obvious that the Kittiwake is very much a gull; innumerable behaviour traits are as similar to the general pattern of gull-behaviour as its morphology and anatomy are similar to those of the other gulls. On the other hand, it differs in many details, of which I have mentioned only a few. The special interest of Mrs Cullen's study lies in the fact that she succeeded in understanding the functions of many of these; she has clearly shown that they must be aspects of adaptive radiation. They are obviously related to the evolution of cliff breeding, which is an extremely successful anti-predator device. Since all other gulls nest on more or less flat ground (even where Herring Gulls or Iceland Gulls nest on cliffs they are rarely on such sheer rocky walls as to be really out of reach of predators), we must assume that this is the more primitive habit in gulls and that the Kittiwake has diverged from the original stock by specializing on cliff breeding. The second interesting conclusion to be drawn from Mrs Cullen's studies is that this specialization has had repercussions on many other behaviour patterns. The Kittiwakes' tameness, its quarrelsome nature, its nest-building habits, the behaviour and coloration of the chicks—these and many other aberrant traits cannot be said to be directly related to each other, but they can all be understood either as necessary corollaries to the cliff-breeding habit or (as in the case of the loss of camouflage) as absence of properties that are indispensable to the other species. Seen from this angle, the Kittiwake stands as a beautiful example of the general rule that adaptation involves the whole animal.

CHAPTER TWELVE

BLACK-HEADED GULLS, 1

Once we had decided to expand our gull studies by comparing different species, our thoughts naturally turned towards one of our commonest species, which was already known, by the work of F. B. Kirkman, to differ considerably from the Herring Gull and the Kittiwake —the Black-headed Gull. It was a stroke of luck that just when we were ready to start on a study of this species, a young American zoologist turned up in Oxford who wanted to do research on the behaviour of birds. Martin Moynihan, a graduate of Princeton University, at once accepted my suggestion that he should study this species. Later, Miss Rita White (now Mrs Weidmann) joined us. While Moynihan concentrated on a study of hostility between individuals and on pair formation, she gave her attention mainly to parent-chick relationships. Still later, Dr Uli Weidmann of Zürich, who had done research in Buldern with Konrad Lorenz, started work on various aspects of incubation behaviour. My own share was again to initiate the work, giving each of them a start, to supervise their first steps and to take photographs. Thus we usually had a party of between two and four camping somewhere near a Black-headed Gull colony.

The first season we went to Scolt Head Island, a narrow ridge of dunes and salt marshes on the coast of Norfolk. Here a small colony of Black-headed Gulls had

settled in the 'twenties on the salt marshes. Thanks to
the hospitality of the National Trust we were allowed
to live in the beautifully situated hut which is available
to field naturalists. The only building on the island, it is
situated on a little plateau on the inner slope of the
dunes, overlooking the saltings.

We shall never forget that wonderful spring on Scolt.
It was fascinating to observe the saltings below, to see the
tide creeping in through the winding creeks until, during
the highest spring tides, the water would rise even over
the vegetation until it stretched all the way from Norton
Creek right up to the foot of our dune ridge. At low tide,
numerous Shelducks, Curlews and a variety of other
waders fed on the marshes. Spring migrants, among them
the sturdy, handsome Ring Ouzels, sometimes alighted
in a tiny Privet bush in front of our dining room.

From the hill just behind the hut we had a wide view
over the sea and on quiet summer evenings we watched
Porpoises and Seals fishing in the shallow waters just
beyond the sand banks. Once we surprised a baby seal
on the shore; Mrs Weidmann's first response was: 'What
a lovely pair of gloves that would make!' but soon her
better self got the upper hand.

However much we liked our first season on Scolt, the
gullery was not particularly suited to our purposes. The
gulls nested on the lowest parts of the saltings and in
order to get at our observation posts we had to cross
several creeks, which made us dependent on the tides.
This defeated our main purpose, yet it gave us some extremely interesting observations on the ill-adapted responses of the gulls to spring tides and flooding.

Some of our observations were made from look-out
posts on ridges from where we could overlook a wide
area. By sitting quietly on the same spot for hours at a
time day after day, we soon tamed the gulls and through
our field glasses, mounted on tripods, we could see a

great deal of interest, even though the vegetation was so high that it often concealed the gulls, or at least all except their necks and heads. For more detailed observations therefore we had to resort to our hides. We used small collapsible canvas tents, 4 ft. cube, with windows in each side. Each window had a flap which could be lowered or lifted. On the outside beneath each window was a row of pockets into which we could put some leafy stalks; while this screen was in bright daylight it concealed whatever was behind it.

Fig. 21. Gull on top of hide.

Because we left these hides standing at the same places for many weeks, the birds became completely indifferent to them; in fact they became too tame and used the hides as look-out posts. This is the most awkward thing that can happen to the observer, since the only place where a bird is absolutely out of sight is on the roof of one's hide. And imagine a male gull on the roof, screaming at the top of his voice within six inches of your ears!

The next season we were allowed to camp on a private estate near a gullery situated on a small island in a freshwater lake in East Anglia. Here our work proceeded very well indeed, but we had the misfortune to see all the young die when they were about three weeks old; and since the owner felt uncomfortable about the possible relation between this calamity and our presence, we felt that we could not possibly risk disturbing this colony for another season. Having spent many years doing this kind of observations in numerous gulleries, I knew only too well that it was highly improbable that our presence had anything to do with this wholesale mortality, but since we could not prove our innocence we were in an awkward position.

From then on we had our base in the huge gullery on the peninsula of Ravenglass in Cumberland. Its owner (Major G. Pennington), the Nature Conservancy and the Cumberland County Council generously allowed us to camp in our tents and a caravan on the southern tip of the line of dunes, and to do all the observations and experiments we wanted. Our party usually travelled up early in spring, the Land Rover bravely towing the 1½ ton caravan all the way from Oxford to Ravenglass and then along the two miles of sandy beach from Drigg to the southern tip. After we once got stuck with our caravan just a few yards below the spring-tide mark near our camping site, and had to be hauled up by an Army DUKW, a local farmer, Mr Jackson, now gives us a hand each year with his tractor.

As on the Inner Farne, our problem in Ravenglass was drinking water. Carrying it over the shallow estuary all the way from the village would have been rather a job, but fortunately sinking a large barrel in the floor of one of the deep wind-kettles near the camp provided us with all the water we needed; the rainwater, which sank into the clean sand, was filtered by it and welled up from the bottom of our barrel. We must acknowledge here the expert help we received from our Hebrew colleague, Mr Amotz Zahavi of Jerusalem, who, when staying with us in 1955, gave us some neat lessons in how to live in a desert.

Camping in Ravenglass was a delight. The sand dunes

were terribly damaged by Rabbits and wind erosion, but such wind-swept sand hills are of a rare, rugged beauty. Since myxomatosis has swept the area, practically eliminating the Rabbits, the vegetation, particularly in the low, moist valleys, has staged a remarkable recovery, showing what such an area could be like in the absence of this introduced, undiscriminating vegetarian. In the distance, the mountains of the Lake District are usually visible. Often, however, they are hidden by low-hanging rain clouds, even when the low coastal strip has relatively clear weather.

The general social organization of a Black-headed Gull colony is very similar to that of Herring Gulls and Kittiwakes. The birds are obviously social, crowded together on relatively small areas even when other suitable nest sites are available all over the peninsula. Within the colony there is a system of territories, which are the properties of individual pairs. The territories are smaller than those of the Herring Gulls, but larger than the ledges of the Kittiwakes. The birds are normally monogamous. Some birds arrive paired, others form pairs after arrival. Both partners incubate and take care of the young. Members of a pair know each other individually and so do neighbours.

Early in spring the birds, gradually returning from their winter quarters, assemble on the fields and shores not far from the colony site. They feed either on the mud flats of the estuary or on the fields, following the ploughs in excited little bands. Much fighting and courtship goes on even before they first visit the colony itself. Settling on the colony grounds is often, though not always, accompanied by the same signs of hesitation, uneasiness and even fear, as in both other species. This uneasiness was much more marked in the inland colony, where the birds had to come down in a restricted area among tall woods, than in the flat open colonies on the sea coasts. In the wooded area a cloud of birds would on successive days show their interest in the breeding grounds by circling above it, gradually descending lower and lower without actually alighting. Now and then they would suddenly stop calling and fly off again, first climbing rapidly until they were once more above tree top level and then sweeping away to the fields all together. On subsequent days they would venture increasingly lower down, but it took them several of these flying visits before they actually alighted on the ground. In Ravenglass the birds, on those occasions when we watched their very first arrival, did not show any hesitation; one day large numbers would suddenly fly towards the sand dunes and just alight.

As soon as the birds have arrived on the breeding grounds, there is much calling and posturing. At first one has not the slightest idea what all this commotion means, but when you follow the birds closely, you see soon that all they do is fight each other, intrude into each others' territories, withdraw when the owner shouts or postures and rush up to strangers when these in turn trespass. Soon one can recognize individual birds and see that there is some regularity in their habits: many return to the same territories day after day, although the whole picture is confused by birds coming one day and never returning, others shifting the location of their territories, and newcomers coming in almost daily. Yet some general trends are clear.

The birds that arrive paired tend to occupy a territory and stay there—that is, return there almost every time they alight in the colony at all. When other birds intrude, the owners, particularly the males, respond by either attacking them outright or by posturing and calling. The actual attacks are rare, but sometimes we could watch prolonged fights. Fighting is done by delivering vigorous pecks with the strong bill, usually from above or, when two birds come to grips, by striking the rival with the folded wings. This is the general gull-method. Black-headed Gulls differ from Herring Gulls in that they are more inclined to fly towards an opponent than to walk up to him.

The various displays are also very similar to those of the Herring Gulls. Moynihan made a very careful analysis of these postures, comparing them with the movements of overt attack and escape, recording whether they alternated with fighting or with other behaviour, and analysing the situations which gave rise to each posture. He found that they must be explained in a way similar to hostile postures in other animals. Briefly, the evidence shows that such postures are the outcome of the simultaneous arousal of two tendencies, that to attack another bird and that to flee from it. Because of this dual motivation, the bird is in a state of internal conflict between these two opposite tendencies and as a result it can neither attack nor escape, but postures and calls.

A very common display is the Oblique Posture, in which the gull utters the 'long call'. A male, upon seeing another gull approach in the air or on the ground, stretches its neck and utters a series of long-drawn, raucous calls. Unlike the Herring Gull, the Black-headed Gull does not bend the neck at the start of the calls and consequently does not show the 'throw-back', but each call is given in the Oblique Posture, although the exact angle of the neck and the degree of curving it slightly down differs from one occasion to the other. Because this display often precedes an actual attack, it must express an aggressive mood. This is also indicated by the raising of the carpal joints, which is always an indication of a tendency to attack. Yet such birds often stop halfway in a charge, particularly if the opponent does not actually intrude, but stays a little outside the territory's boundaries. This halting halfway during a charge also shows that the posture is not the outcome of pure aggressiveness, but that fear is also involved. Finally this is indicated by the fact that the birds often alternate between rushing up towards an opponent and withdrawing.

Another posture which occurs in hostile encounters is the Upright Threat Posture. It is very similar to that of the Herring Gull (the Kittiwake does not have it), but the carpal joints are usually raised much more conspicuously than in the Herring Gull. When a bird follows this posture by an attack (as it often does), the wings are moved up and all the intermediates between just lifting the carpal joints and actually spreading the wings and flying towards the intruder can be observed. In interpreting the motivation underlying this posture, therefore, we use an additional criterion—the actual form of the posture. Comparing it with overt attack, one sees at once that it involves parts of it. Yet a bird in the Upright Posture does not always attack, but more often than not stops when it comes near the opponent; this happens almost always when the latter does not give way at once. It is just on such occasions, when the attacker is obviously held back by fear, that one sees the most intensive posturing; the neck is swollen and the carpal joints raised very distinctly. Both the form of this posture and the underlying motivation could be compared with that of an enraged man who, keeping himself under control (or being simply afraid), clenches his fists.

A third posture—very common in this species—which is shown in hostile situations is the Forward Posture. The bird stands in a horizontal stance, often facing its opponent, and points its beak forward. It is very similar to the incipient biting movement observed, though not as frequently, in Herring Gulls and in Kittiwakes and which, in the Kittiwake, we call 'jabbing'. Although in the Black-headed Gull the posture is not often correlated with actual biting, its close association with hostile clashes and its similarity with jabbing leave little doubt that it is also the outcome of the simultaneous arousal of attack and escape tendencies.

The fourth hostile posture is again one which we knew already in other gulls—Choking. A 'choking' gull bends down, points its bill at the ground and, with curiously muffled rhythmic sounds, makes pecking movements towards the ground, although it rarely touches it. Its derivation is not completely clear, but there are good indications, most obvious in the Kittiwake, that it is a displacement activity derived from the movement of depositing nest material. According to our present views, this movement is also the outcome of an internal conflict between attack and escape, and seems to indicate a very high level of both.

It would carry me too far to discuss these postures in more detail. It does seem as if the interpretation I have given here is roughly correct, but we are certain that we have not yet found out the whole story. For instance, each of these postures can be shown in slightly different forms and we believe that the relative intensity of the two underlying urges controls part of these variations. Thus a bird in the Oblique bends its neck back and points its bill up when it becomes more afraid. The Forward Posture seems to be different in more aggressive than in more timid birds; in the latter the bill tends to point up.

It will take several more seasons of careful study to disentangle the motivation underlying these postures in more detail, and we may still be in for some surprises.

All these postures have obvious signal value; they convey a message of some kind to fellow Black-headed Gulls. This is clear from the responses shown by them. For instance, as soon as a bird begins to walk towards another bird in the Upright Threat Posture, the latter begins to sidle away. Birds even understand whether the posture is aimed at them or at others; often it is the one bird towards which the threatening bird turns who shows concern, while all others ignore the whole thing, even though the threatening bird may pass within a foot of them. Numerous similar 'natural experiments' do not leave much doubt about the intimidating effect of the threat postures: birds often either flee, or are stopped in an incipient attack, when met by counter-threat. On the other hand, under certain circumstances a threat posture may draw the wrath of another bird towards the threatening bird; this provoking effect is seen when an intruder happens to adopt a threat posture while he is well within the territory of another bird—a kind of mistake which can occur in confused situations when several birds are involved in skirmishes.

We would have liked to have more exact experimental evidence of these effects, but we found it extremely difficult to obtain. The provoking effect of a stuffed bird put in a territory is easily seen; such mounts are often vigorously attacked even if they are mounted in a neutral posture. We hoped it would be easy to study the difference in vigour of the attack correlated with the posture in which the dummy was mounted, but in practice Mrs Weidmann found that large numbers of such tests, required for valid conclusions, would take a large number of models, the dummies soon being damaged by the gulls' attacks. Once an attack had begun, nothing could be done to stop it, short of running out of the hide from which we were watching, and this would soon scare the gulls away completely; we simply had to wait until an outside helper, whom the gulls were used to, would relieve us. Before he could get there, however, the damage was usually done; whole dummies were torn apart in a few tests.

But we did see some rather suggestive things. For instance, a dummy mounted in the Forward Posture was once attacked vigorously by a male. When he pecked at it and beat it up, it lost its balance and leaned backward, thus pointing the bill up in a way which reminded us of a female's Hunched Posture, adopted when she is begging for food. When this happened, the male stopped his attacks and his angry calls, suddenly bent down and, regurgitating in the way males do when feeding their mates, produced a large earthworm! However, our present evidence does not go beyond such occasional results and we are still trying to develop better dummies.

One thing has already been well established by tests of this kind: attacks are provoked by the head of an intruder rather than by anything else. Huxley and Fisher showed that just a head of a gull mounted near a nest was furiously attacked. Mrs Weidmann found the same; we have some convincing film scenes showing a concerted attack by three neighbouring gulls on a head alone —and curious lack of concern by the same gulls when a body without a head was put up in exactly the same spot. Neither Huxley and Fisher nor we could see a noticeable difference in the response towards brown and white heads, yet I believe that such a difference exists, though it seems to be slight.

We often wondered why a species would have so many different threat postures. The solution may have to do with the fact, which seems fairly well established for some of these postures, that they express different things. The Upright Threat Posture, for instance, seems to be an expression of a relatively strong and little inhibited tendency to attack, whereas 'choking' is performed by birds which at the moment are more inhibited by fear, but which are prepared to put up a fight only if approached too closely. Also it seems that some postures, such as the Oblique, are due to a lower absolute level of attack-escape conflict than others such as 'choking'. Although we cannot claim to understand this problem fully, it seems that these absolute and relative intensities of arousal have much to do with it. Also, the opponents respond differently to the different postures and in, roughly, a suitable way. There can be little doubt that the general conclusion is correct, namely that these postures are first and foremost hostile and are due to both attack and escape tendencies.

In these hostile encounters one sees a number of other activities, which have obviously to do with the state of tension the birds are in, but of which we do not know whether they have any effect on the other birds. Sometimes an angry bird pecks violently into the ground. It may sound fantastic, but it rather looks, judging from the type of movement, that he is indeed attacking the ground and it may well be that this is an extreme case of what is called in technical language a 'redirected attack'. It occurs when a bird is strongly provoked by another, but is inhibited from attacking it. Less extreme cases of this are known to everyone, because it is a common feature of human behaviour: when a man has been ticked off by one of his superiors at the office, he is likely to be, let us say, less friendly to his wife when he comes home. Nor are we human beings less liable to do such a stupid thing as 'attacking the ground'; who has not seen a man kicking a chair or a door, or throwing his razor or a book across the room when extremely angry, when the object arousing his anger may have been one of his fellow men?

Another sign of tension seen in Black-headed Gulls is preening. It usually takes the form of some stereotyped, rigid preening movements of either the flank feathers or the scapulars.

In recent years we have also paid some more attention to the movement of turning away the face ('head flagging'). We noticed it first in courtship, where it is so conspicuous that one cannot miss it, and which I will discuss presently. We thought that it was peculiar to courtship and to this particular species. But we have since seen that it occurs in Kittiwakes (where it is common in fights and is shown by the beaten bird) and in Herring Gulls. Also we noticed that Black-headed Gulls often do it before or during fights. It seems that in all these species head flagging is the outcome of a tendency to flee, but that complete overt withdrawal does not follow because the bird is at the same time compelled to stay. In Kittiwakes it is a means of protection employed by a non-attacking bird, but Black-headed Gulls may do it even while rushing up to attack another bird. At the moment not much more can be said; it is one of those things which simply have to be studied in more detail.

Moynihan spent much time in observing pair formation. This can be observed all through the breeding season, though naturally more frequently in the beginning. On Scolt Head the spring tides, which flooded all the nests, forced the birds to start a new cycle more than once, for each cycle many pairs being formed anew. In all colonies, young birds come in during June or even July, and these birds may go through part or all of the pair-formation ceremonies.

It all starts with the males taking up small temporary territories, sometimes in the colony, often near its borders, so-called pre-territories. Such males are extremely alert and continually scan their surroundings. Whenever another gull passes within about 15–20 yards, he responds by the 'long call'. Since there are always hundreds of gulls flying about in a large gullery, and since the 'long call' is also uttered by any bird resenting near-intrusion, a Black-headed Gull colony is in continuous uproar. The din is terrific—incidentally a boon to us, since there was no need at all to be silent in the hides; two observers could talk without the slightest fear of disturbing the birds and the buzz of the ciné-camera was also completely drowned.

Now among the birds cruising round over the colony there are always several unmated females. Unlike the aggressive males, who, when on the wing, make strong, exaggerated wing beats and chase other birds almost continuously, the females fly in a quiet, subdued way, with shallow wing beats. Now and then a female alights near a calling male. And then we saw a spectacle which puzzled us a great deal at first, until we began to see that it was a special case of a very wide-spread phenomenon. Both male and female adopt the Forward Posture, remain in it for some seconds, and then first one, then the other, suddenly jerks into the Upright Posture and, with an equally startling jerk, both turn their faces away from the other.

Often there is a slight but significant difference in the postures of the male and the female: while his neck is swollen and his bill points down, she makes herself extremely thin and sleek and often points her bill slightly up. Both birds raise their carpal joints. Their postures are very similar to the Upright, but that adopted by the female shows relatively more signs of fear than that of the male. Judging from the behaviour then we would have to conclude that the male's predominant condition when calling before the female alights is aggressive; that the female is attracted to the male; that after alighting both birds are motivated by a mixture of aggression and escape, which persists when they adopt the Upright Posture, but in which an element of fear makes them show the 'head flagging'. This behaviour may seem surprising in birds that come together for the purpose of mating, yet there are other indications that they are in a state of extreme tension, in which aggression and fear mix with sexual motivation. For instance, at this stage the male often pecks the female away. Even if he does not show any overt aggression towards her, her presence often causes him to become much more intolerant towards his neighbours. Or her presence may make him peck violently into the ground. Signs of the female's fear are only too obvious. At the slightest movement of the male she jumps or at least turns her face away again. Very often, at this early stage, she flies off soon after she has landed.

Furthermore, from what we know of the stimuli that release hostile tendencies in the male, it is not really astonishing that the female arouses his hostility; it is even to be expected. Hostility flares up at once whenever another Black-headed Gull trespasses into his territory and since the female's plumage is so very similar to that of the male (it takes an experienced observer to tell the sexes apart), she cannot possibly avoid providing some of the stimuli that elicit a hostile or partly hostile response.

I regard this result of our analysis of threat and courtship postures as one of the most interesting aspects of our work. It throws light on the origin and significance of all these elaborate courtship displays and makes it understandable why such displays have evolved at all. Hostility plays such an important part in the spacing-out of breeding pairs that the species seems to be unable to give it up. And because it is elicited by stimuli provided by the intruders, the males seem to be unable to pair straight away, but are thrown into a state of internal conflict as soon as the female approaches them.

Because of this it seems that the 'head flagging' has a very important function indeed. Its effect on a fellow member of the same species is very striking: as soon as a bird turns its face away from its partner or its opponent, the latter is practically unable to attack. This effect is very conspicuous in fights between Kittiwakes, but in the Black-headed Gull it can also be seen. The peculiar thing about gestures such as 'head flagging' is that they do not suppress attack by scaring the other bird; they seem to have a soothing, an appeasing effect. This is why we call 'head flagging' an 'appeasement ceremony'. Such a ceremony is important just because hostile tendencies occur in courtship; an appeasement gesture is exactly what is needed when the prospective mates still distrust each other and respond to each other's presence by threatening. How it has come about that a posture, which is nothing but the outcome of the simultaneous arousal of fear and a tendency to stay, has acquired this appeasement function is still an open question.

Courtship of other kinds of animals have been investigated with this idea in mind. In finches and other birds, in fish such as sticklebacks, the River Bullhead, Guppies, and Cichlids there are many indications that this conflict theory of courtship is fundamentally correct. However, it works out very differently in different species. It would lead me too far to discuss the problem in more detail here.

If this is the explanation of the Black-headed Gull's courtship, then how do the birds succeed in mating at all? Continued observation of pair formation behaviour provides the answer. In the beginning, a female leaves the male soon after she has alighted near him. She may then visit other males (sometimes one after another in quick succession) or she may return to the same male. It seems that one female may fall in love at first sight, while others flirt with several males before finally making up their minds. When a female becomes more or less attached to one particular male, she visits him again and again, every time both birds going through the full ceremony: Forward, Upright and 'head flagging'. But the character of this greeting ceremony changes. The signs of fear in the female subside. Her plumage is no longer sleeked so much, her neck is no longer so stretched, she comes closer to the male and she is less inclined to flee. The male's aggressiveness remains, but he is less likely to peck at her; instead he vents his aggressive tendencies in attacks on other males. Once the birds are really paired, they know each other individually and it seems likely that this gradual reduction of the tension, of hostility and distrust, is the consequence of the two partners getting used to each other. Repeated approach and repeated appeasement by 'head flagging' seem to be the means by which this is achieved.

Once this stage is reached, the development is very similar to that in other species of gulls. The female begins to beg for food and the male responds by feeding her. Soon the birds copulate, this being introduced, as in other gulls, by mutual head-tossing, the same movement as that used by the female when food-begging.

Now the pair abandons the pre-territory—at least if this had been situated outside the colony. Together they go 'house-hunting'—i.e., they visit the colony proper and find a suitable, non-occupied territory, where they settle and which they now defend against intruders. This stage often involves a great deal of fighting and hostile shouting, since the boundaries with already occupied territories have to be established.

There are still a number of aspects of this story which we do not yet understand fully. We are still far from being able to predict in any type of encounter which posture the birds will adopt—which means that we do not yet fully understand the causation of the postures. We are searching for ways to study some of these aspects experimentally; for instance, more tests with dummies and with mirrors are planned. We don't know how illuminating they will be. Some of my own co-workers are doubtful of our chances of success, but I believe that it is worth trying. So far each new season has carried us a little further. Usually it pays to stick to research once begun, and to try to design new methods of study.

CHAPTER THIRTEEN

BLACK-HEADED GULLS, 2

As I have already mentioned, the original aim of our observations on Black-headed Gulls was a comparison of their behaviour and postures with those of other gulls, to find out something about the evolution of such displays. This work has not yet been completed. However, while we were doing it, we found that Ravenglass offered excellent opportunities for studies of a variety of other problems as well. And, having already invested in such costly and, in part, specialized equipment as a car and a caravan, drinking-water barrels, tents and Primus stoves, canvas hides, and many other things, we were naturally reluctant to stop our Ravenglass programme just when we began to realize how much more we wanted to know. So we extended our programme and so far have not regretted it. Though I have to admit that our progress has been slow, some of the results are already worth mentioning.

The reproductive cycle of Black-headed Gulls, like that of so many other animals, is a succession of several phases. The birds start by becoming aggressive; then they begin to form pairs; this is followed by a period of mainly sexual behaviour; after this the birds build a nest; they then start to incubate; and when the eggs hatch the parents guard and feed the chicks. On the whole, as little is known about the causes of these major shifts in the birds' behaviour, the Weidmanns thought that the Ravenglass gulls might be good objects for a study of such shifts.

Some of their observations concern the onset of incubation. From the fact that gulls are not at all times ready to incubate eggs it is clear that the start of incubation is partly dependent on an internal change, on the bird becoming 'broody'. Further, apart from broodiness, stimulation by the sight of eggs in a nest is required to make a bird brood.

We first decided to examine the influence of external stimulation and started on two aspects. By offering broody birds a choice between different models of eggs—with varying colour, size, shape, etc.—and recording their responses to them, it was possible to analyse the stimuli that, to a gull, characterize an egg. This analysis was begun by Moynihan. It proved to be extremely time-consuming and, since this was a side line to his main research, it has not yet been carried far enough. One striking result was obtained: just like Herring Gulls and Oyster-catchers, Black-headed Gulls strongly prefer eggs of grossly supernormal size; they always preferred eggs twice the linear size (8 times the volume) of their own eggs and would make desperate attempts to sit on them. Many other aspects of 'egg recognition' were tentatively explored. The whole problem proved to be much more complicated than we had expected and the only thing we can say at the moment is that we hope to tackle it more thoroughly later on.

The Weidmanns began to study broodiness and their first step was to offer normal eggs at various times during the reproductive cycle—before the birds had themselves laid eggs; while they were sitting on their own clutches; and after their own eggs would normally have hatched. One difficulty of this programme was that gulls which are not broody do not always just ignore eggs, but often eat them. This is why wooden egg models were added to our equipment. Good imitations of Black-headed Gulls' eggs were mass-produced in the lab. —an occupation which evidently appeared a little frivolous to some of our colleagues. The gulls accepted them readily when they were given instead of their own eggs. Clutches of wooden eggs were put in empty scrapes of which the owners were known and the responses of the owners were regularly checked till the moment they laid their own eggs. Of course, it was impossible to know in advance when such birds would lay their own eggs and we could not, therefore, plan a neat series of equal numbers of tests running from long to short intervals between the tests and the appearance of eggs.

Many birds sat on the prematurely offered eggs. Some were willing to sit a full fortnight before they laid an egg of their own, others not until a few days before their own egg was due. It was also found that gulls are prepared to sit much longer than the normal incubation period—a fact also known in many other birds.

The readiness to sit on eggs before the normal start of incubation was present in males as well as in females, who in all gulls, take about equal shares in incubation. There was a wide individual variation. Some birds incubated 18 days before laying, but several others (3 out of 8 birds tested) still ignored eggs less than 4 days before laying. As usual, things proved to be complicated; for instance, several birds turned the eggs though they did not sit down on them; many birds pecked at the models before sitting down or ignoring them; if they had been offered real eggs, they would have damaged them and almost certainly would then have eaten them. This pecking indicates incomplete broodiness, for gulls do not peck at their own eggs. Either the act of laying itself abolishes this tendency, or pecking disappears shortly before laying—at least as far as a bird's own eggs are concerned.

These tests, mere preliminaries to a serious study of broodiness, also gave information of quite another nature. Some of the gulls, which had accepted wooden eggs long before they were due to lay eggs themselves, never laid any eggs at all! This happened with so many of them that it was impossible to assume that none of these birds would have laid eggs anyway—they must have been stopped from developing eggs at all by the presence of the wooden eggs. This was unexpected, for gulls had usually been considered to be 'determinate layers' —i.e., species in which the size of the clutch was internally determined and which could not be forced to lay either fewer or more eggs than usual by giving or, alternatively, taking away eggs. Other species can be 'milked' almost indefinitely by taking away eggs soon after they are laid.

Although two other observers (Paludan and Salomonsen) had already shown that Herring Gulls are not so rigidly determinate layers, but could be influenced, this was worth corroborating and expanding. Weidmann soon found that under certain circumstances the removal of eggs could force the gulls to lay one or even more extra eggs. This was possible when the first egg was removed soon after it was laid or when the first egg was left but the second was taken away promptly. But when the gulls were allowed to keep the first two eggs and only the third was taken away, no extra egg was laid. Yet a bird could be made to lay six or even seven eggs in regular succession if every egg was removed soon after it had been laid.

It was clear, therefore, that a gull can lay considerably more than the three eggs it usually lays. Inspection of the ovaries of females when they were due to lay their first egg showed that there were at least four eggs well on their way, with many more available and already slightly developed. Normally, however, the fourth egg degenerates again before it is laid. But in some of the tests this fourth, and even subsequent eggs, developed and were laid. There must, therefore, be something in a normal gull that stops this process. This could be due to stimuli (either visual or tactile) given by the complete clutch of three. But in some tests the presence of one egg was already enough to stop a bird laying!

Weidmann suggests that the inhibiting influence is not directly dependent on the eggs, but on the onset of broodiness. Broodiness develops under the continued influence of eggs; if one egg is left in the nest long enough this will make the bird broody. The onset is gradual; normally gulls begin to incubate regularly at about the time the second egg is laid, or a little before, and this may explain why a bird continues laying when each egg is taken away soon after it is laid and why, if the first egg is not removed though the second is, some birds lay a fourth egg while others don't; this would depend on the amount of brooding the bird had already done when the second egg was taken. Thus, ultimately, the end of laying is dependent, in part, on the presence of at least one egg. The effect is indirect: an egg brings about the onset of broodiness, which stops the development of eggs. And this also explains why those birds which were given wooden eggs before they had laid themselves never laid at all: they were made broody sufficiently long before they would have laid to make even their first egg degenerate.

Thus these facts show the influence of the presence of eggs on broodiness and, beyond this, on the development and subsequent laying of eggs as well. They also gave us a glimpse of those curious relationships between behaviour and 'somatic' processes such as growth phenomena in the ovaries. How these things are controlled in the males remains an open question.

Broodiness entered into our studies in still another way. We had often observed, at first without attaching too much significance to it, that gulls (and other birds) often do some or even much nest building long after the nest has been constructed and the eggs have been laid. Gradually it began to dawn upon us that there seemed to be some connection between broodiness and this seemingly irrelevant nest building. When we began to record exactly when an incubating bird, or at least a bird in the incubation period, did this nest building, we found that it usually occurred in the following situations: (1) when a bird tried to relieve its mate but the mate refused to budge from the eggs; (2) just after a bird had been relieved; (3) just before a bird got up from the eggs in order to shift them; (4) when a sitting bird saw its mate arrive after a prolonged absence; and (5) when the eggs had hatched and the young chicks crawled out from under the stitting parent.

In experiments with egg models, we had also noticed that birds would often start making building movements when they were sitting on abnormally shaped egg models or on empty nests. All this together seemed to us to suggest that nest building could appear as a 'nervous outlet', a displacement activity, when a bird was broody and yet in some way frustrated—e.g., by its mate not allowing it to sit or by sitting on unsatisfactory eggs.

There were various possible methods by which this could be approached experimentally and many reasons why it should. Moynihan made a start by carrying out a series of tests in which he watched the nest building activities of gulls sitting on the normal clutch of 3 eggs and of gulls whom he had robbed of either 1, 2 or all 3 eggs. He found that nest building became more frequent, and more intensive, the more eggs were missing.

Birds sitting on empty nests even went on special collecting trips and carried in quite a lot of material; there was
a close correlation between frequency and intensity of
nest building and the number of eggs missing. It looked
as if the stimuli that the bird receives from the eggs on
which it sits have to 'signal back' the information that
the clutch is complete and normal; the less complete this
message is, the more intense is the frustration. Why the
bird should be frustrated in this situation—i.e., whether
it is the direct effect of the abnormal stimuli or whether
the abnormal stimuli cause anxiety which then comes in
conflict with broodiness—is still an open question, although several observations point to the first conclusion.

An extremely interesting point is that it cannot be due
to the bird being prevented from making the incubation
movements. Even on an empty nest it sits in the normal
brooding position. Further, it is not clear why frustration
of this type should give rise to building movements and
not to other behaviour. Moynihan's tests mark a first step,
however, and are well worth continuing, for it is still
largely obscure generally what causes frustration, and its
counterpart—satisfaction.

All such studies meet with one serious handicap: the
breeding season is short and each year allows one no
more than a few weeks for any particular problem. However, the total reproductive cycle takes months to
complete, and since each phase offers plenty of interesting problems, we switch, with the birds, from one
phase to the next, adding a little to our results every
year.

Thus the time comes each year when the eggs hatch.
This initiates a change-over to parental behaviour. By
presenting chicks too early or, alternatively, too late, the
part played by external circumstances and the possibility
of an internal change can be studied in the same way

as the onset of incubation. So far, Mrs Weidmann has done only a small number of tests of this type. She offered downy chicks to incubating birds at various stages of the incubation period. The surprising result was that they were almost always accepted; some chicks that were given less than ten days after the laying of the first egg were rejected. One bird accepted (i.e., brooded and fed) a chick on the day after laying its first egg. Mr Zahavi even saw a gull accept a chick before the first egg was laid!

Yet this does not mean that a bird which is broody will always accept a chick. First, several incubating birds rejected chicks and, second, parental care and incubation seem to be mutually exclusive to a certain extent. At any rate, a gull which accepts a chick usually abandons its eggs within a few days at the most.

With the appearance of chicks another opportunity for field experiments presented itself. Very soon after hatching, sometimes before the downy plumage was dry, the chicks began to beg for food. They stood up in the nest and pecked at the bill tip of their father or mother—whoever happened to be near. This soon made the parent regurgitate food, which the chicks then picked up. It was obvious from the beginning that this begging behaviour was a response to visual stimuli provided by the parents.

Years ago I had, together with A. C. Perdeck, studied this behaviour in chicks of Herring Gulls by presenting them with models which were visual imitations of the parent's head and beak. We had found that the chicks responded very well indeed to these models. This had been the starting point for many tests in which we compared the effect (in terms of pecks per 30 seconds) of different models. By varying colour, shape, type of movement, distance, level and position of the model we found out, step by step, to which stimuli the chicks actually responded and which part of the environment did not affect them.

Some of the results seemed rather striking. For instance, the chicks did not respond to the parent as a whole but only to its bill. It did not make the slightest difference whether the model was flat or solid. The red spot on the lower mandible of the parent provided a particularly powerful stimulus, and it acted by its colour as well as by its contrast to the yellow colour of the rest of the bill. It did not matter in the least, however, what colour the bill had as long as it showed this red patch. The shape of the bill was also important: it had to be thin and long and had to point down—a horizontal bill gave only very few responses. These results were the more remarkable since in some tests we used chicks that could not possibly have had any experience of what the parents look like, and from the few tests we did with such inexperienced chicks it seemed as if the response was unlearnt.

For various reasons we wanted to repeat such tests with Black-headed Gulls. First of all, because the bill of this species is entirely red and not yellow with a red spot at the tip. Further, we wanted to do this kind of test more accurately than Perdeck and I had done, also, it would be of great interest to find out exactly what stimuli the chicks responded to before any learning, for surprisingly little is known about the total stimulus situations eliciting non-learnt behaviour. The enormous size of the Ravenglass colony promised us large numbers of chicks to work with. Mrs Weidmann, therefore, selected this as a subject for her doctor's thesis and spent four seasons in the field doing large series of model tests. She succeeded in working out a detailed description of the total effective stimulus situation and also improved the experimental technique.

In order to get chicks that had never seen an adult gull when they were subjected to these tests, Mrs Weidmann collected eggs which were just chipping, took them home and hatched them herself. At first she did this by packing them in a cardboard box and taking them into her sleeping bag for the night. Next morning, after a none too comfortable night, she could proudly produce a box of chicks. Later she used an incubator; this provided better conditions, safeguarded her well-deserved nocturnal rest and gave the chicks still less chance of getting acquainted with their normal environment. Collecting the eggs, checking the incubator several times each night, doing the experiments and returning the chicks to the nests where they belonged (and where the parents accepted them without ado) kept her and her husband pretty busy all through the short season when the eggs hatched. As in all the other seasonal work, there was always a race with time.

In the tests, a chick was shown a model for 30 seconds and the number of pecks it aimed at the model was registered. Great care was taken to standardize the tests as much as possible: distance between model and chick, the model's position, the type of movement, the background, etc. were kept constant unless their influence was itself investigated. Each chick was presented with all the models of a series in the course of a test series, but the sequence in which the models were presented was varied from chick to chick, according to a prearranged plan.

In all, 503 chicks were used for 2,431 tests. The effectiveness of each model can be indicated by the average number of pecks it released in comparison with those elicited in the same series by a standard model, which was a flat cardboard imitation of the front part of the parent—bill, head and neck in sideview, in approximately the natural colours. I will give these averages here only when statistical treatment showed them to be significantly different from the average obtained with the standard model.

When the models were moved slightly from side to side in front of the chicks, they elicited more pecks than when they were kept still (10.3 versus 3.4). Therefore, in all other tests the models were moved in this way.

Models which differed in the colour of the bills were presented first. As we expected, red bills elicited more responses than most other colours or grey bills, with the exception of a dark blue bill, which was almost as effective as red. The fact that not all blue bills had this effect worried Mrs Weidmann a great deal. Measuring the reflection of the papers for the visible range of the spectrum (which is probably the same for gulls as for us) supplied the most likely answer: this particular paper reflected a great deal of red as well as blue and since most birds are a little less sensitive to blue than we are, this paper might well have been redder to the bird than to us.

Long series were done in which the normal brown colour of the face was varied; but the chicks responded equally well to all of these. The same result was obtained when the white colour of the models' 'neck' was varied while everything else remained constant.

The shape of the head could also be varied without affecting the chicks' responses in the least. Finally, when Mrs Weidmann compared the effect of a complete standard model with that of a bill alone, she found no difference. This was not as astonishing as it seemed at first, because there were several indications that the chicks' field of vision was relatively small and that they simply did not see much more than just about the bills of the models; it was true that they would peck now and then at other parts, but they did this only when they happened to have turned their head and so were accidentally facing another part of the model.

The rest of the tests, therefore, were designed to analyse the properties of the bill which released the begging. The part played by movement and by colour had
already been studied. When one and the same model
was presented at three different levels (high, which
was 2 cm. above the eye level of the chick; medium, with
the tip of the bill at eye level; and low, with the tip ½
cm. from the ground—which was several cm. below
the chick's eye level) the high position proved to be in-
effective, while low and medium were only slightly, if
at all, different (high 3.7, medium 16.3, low 19.9). This
again may be due mainly to the chicks often failing to
see the high model at all.

The thickness of the bill proved to be relevant too:
a bill 16 mm. thick received 13.8; one of the natural
thickness of 7 mm. elicited 16.4; and one of less than
half that thickness (3 mm.) scored 17.8. Whether the
last figure was really different from 16.4 is doubtful, but
the thick bill was significantly less effective than the
thinnest model. When a bill of normal length was compared with one more than double its length, there was
no difference in response.

However, the position in which the bill was held
(vertical, oblique, horizontal) was important to the
chick. Horizontal scored 12.2, oblique 18.8, vertical
19.2. In another, longer series especially intended to
compare oblique with vertical, the scores were: 17.8 for
oblique, 23.6 for vertical. These tests proved that vertical was better than oblique and oblique better than
horizontal.

Since we had found previously that the chicks of
Herring Gulls were stimulated strongly by the presence
of a contrasting spot on the tip of the parent's bill
(which corresponded to the natural situation there, since
Herring Gulls have a red spot at the tip of the yellow
bill), Mrs Weidmann wanted to know whether this

preference was absent in the Black-headed Gull which, as an adult, has a uniformly red bill. No such preference would be expected if the chicks' responses were strictly adapted to the natural situation. To her surprise she found that her chicks were just as fond of such contrasts as Herring Gull chicks: although white bills and black bills gave far lower scores than red bills, a white bill adorned with a black spot was twice as effective as a uniformly red bill (16.3 versus 8.6). Thus the stimulus situation to which the chicks respond best does not fit the natural situation.

This was an extremely interesting and challenging result. First of all, it raises the evolutionary question why this preference of the chick for a contrasting spot has not exerted sufficient selection pressure on the species to make it produce a spot on the bill. This is, so to speak, within the power of the species, for first-year Black-headed Gulls have a pale pinkish bill with a darker tip. I am inclined to believe that the red bill plays a part as a signal structure in hostile behaviour, but we have no proof of this. If this is correct, it might well be that selection pressure in favour of an all-red bill has been stronger than that in favour of a spotted bill. However, this is all speculation.

A second problem concerns the physiological machinery which causes this preference for contrast in the chick. At the moment we can only guess. Some of the limitations of the effective stimulus situation are doubtlessly due to limitations of the eyes themselves; for instance the chick's eyes simply cannot see the greater part of the parent when it is as near as it is when feeding the chick. On the other hand, the preference for red may well be due to the central nervous mechanisms involved, since there are indications that the chicks can see and distinguish other colours as well. However, the evidence here is still incomplete.

One interesting method of checking this was discovered rather by accident. When, as in all these tests, you offer a model repeatedly without ever rewarding the chick by giving it food, the chick responds less and less. This reduction in the number of pecks is partly due to a general effect which recovers in a short time; after a brief rest the chick responds better again. But another part of the reduction is a learning process which persists. And this learning-not-to-peck is very specific: if you show a chick a red model throughout a series of tests and then, after a couple of hours, subject it to tests with the whole range of colours and greys used in the main series of bill colour, it will respond normally to all the models, but show a very low response to red. By doing this kind of tests with colours other than red as well, it will be possible to find out in a short time which colours the chicks can distinguish from each other, and from greys. With the necessary controls this can be used to study colour vision of the chicks. It is also worth investigating as an example of a very quick learning process in a young bird. The Weidmanns are planning to continue this work for a number of seasons to come.

This aspect of parent-chick relationships was of great interest to us because it demonstrated not only how a very specific stimulus situation affects the chick even before it can have learnt anything about its parents, but also how the responsiveness of the chick can be changed by conditioning. I have already mentioned the effect of 'discouraging', by which the chick learns not to respond to certain stimuli, even such a powerful one as the red colour. There are also indications that the fact that the parent rewards the begging of the chick by feeding it leads within a few days to the chick's becoming positively conditioned to its own parents. This soon enables it to recognize them as individuals.

Mrs Weidmann also did some tests to find out whether

the parents learn to distinguish their own chicks from strangers. I had myself, in doing exchange tests with young Herring Gulls, found that, when chicks are five days old or more, they were no longer accepted by gulls not their own parents. I believed that this refusal to accept strange chicks was not due to the chicks behaving abnormally when faced with strange adults and then being refused because of their abnormal behaviour, but that parents really recognize their own chicks in any circumstances. This is difficult to decide, since chicks usually show that they are ill at ease with strangers. However, in some tests I could not see any evidence of this, yet the chicks were pecked away by the strange adults.

When Mrs Weidmann repeated these tests with Black-headed Gulls, she got rather similar results. When she exchanged the chicks of two nests while both broods were very young—one or two days old—both pairs of parents accepted the strange young. When she did the same with chicks seven days old, the adults often did not accept the strange chicks but pecked at them. However, in most of these tests the chicks did give some signs of uneasiness even before they were attacked and in only a few tests the parents seemed to react before the chicks did anything unusual. No conclusion can be drawn, therefore, but we may solve the problem by doing more tests.

When the young are about half-grown, Black-headed Gull families often leave their territories and wander off. This causes much commotion, since these moving families trespass continually on occupied territories, and violent clashes result. To our surprise not only the adults take part in such fights, but the young as well. This provides good opportunities for observations on the development of hostile posturing in the young birds. These, however, have not yet been completed.

Thus our attempts at understanding the organization of a colony of gulls have led us to a variety of researches. Most of these are still far from their conclusion and I have hesitated a long time before deciding to include the three gull chapters in this book. Some readers may find these chapters too heterogeneous, too much the outcome of opportunism. I finally decided to give them because they demonstrate both the limitations and the positive possibilities of field work on animal behaviour. Frankly I consider our gull studies as a kind of test case of the possibilities of field work and I believe that, in spite of many shortcomings, they show that there are things which can be better studied in this than in any other way. It would seem to be more efficient to try to improve the field methods than to try and keep a large colony of gulls under laboratory conditions. In fact, a colony in or near the lab. would not be so very different from a colony in the field; the main difference would be that the lab. colony would be extremely expensive to run. The use of hides and field glasses, of colour-ringing and thus following individual case-histories, etc. can overcome most of the disadvantages of working in the field.

CHAPTER FOURTEEN

EIDER INVASION

You can't believe that Eider Ducks are true till you see them. The colours of the drake are of a distinguished boldness unsurpassed in the animal kingdom. Jet-black parts are set off sharply from pure white in such a way as almost to conceal the bird's real form. A slight touch of green in the nape and a delicate peach hue on the breast add life to what otherwise would be rather a commonplace dress. As it is, the colour scheme of the Eider drake is of a rare distinction.

The duck, by contrast, has what might seem at first a much simpler pattern. This is true only in so far as she shows no brilliant colour contrasts. But seen from nearby her seemingly dull plumage reveals lovely patterns of bars, waves and mottling in various shades of havana-brown, which are no less delightful than the flashing colours of the male. And far from breaking up the body's outline the duck's plumage brings out the wonderful shape of this sturdy bird. The strong, almost squat body, the graceful neck, the Greek profile, the whole shape of the powerful bill forming one architectural whole with the skull—it all has style.

Eider ducks are true marine birds. They get all their food from the sea. Luckily for them their requirements do not really clash with our own and so fishermen look upon them with tolerance.

There is perhaps no better place for observing Eider ducks than the Inner Farne. We often watched them browsing in the seaweed along the shores of the Kettle, the shallow bay surrounded by our island, Knox's Reef and the Wideopens. Lazily paddling along, they stretched their necks down as deep as they could and took their toll from shellfish, small crustaceans and perhaps worms and fish as well.

But this is not when Eiders are at their best. To get an impression of their real capacities you have to see them diving in deep water. We saw them at work daily at the foot of the Kittiwake cliff. They were so used to our almost continuous presence at the top that they completely ignored us, except for occasionally cocking their heads and looking up at us with one eye. When the sea was not too rough they came right up to the foot of the cliff. Every now and then one gave a powerful push with the two strong feet and started on the way down. Just before they went under they opened their wings and then, if the weather were calm and the sea smooth, we could enjoy the marvellous spectacle of the ducks flying under water. The dark females disappeared from sight almost at once, but the pied males could be followed several feet down, for the water was wonderfully clear. With each wing beat the white bird grew dimmer and dimmer, until it was lost in the green depths.

Sometimes, however, the ducks travelled near the surface and then we could follow them for up to a minute at a stretch and see them explore the rocks, leisurely 'flying' along the submerged cliff. It was fun to see them come up. They just half-folded their wings and shot up buoyantly, breaking the surface like a ping-pong ball released under water. Their usual food seemed to be Limpets, but we also saw them come up with other shellfish, for instance the long Razor Shells, and it was truly amazing to see them swallow specimens of 4 to 5 inches long!

We sometimes saw these travel down as curious bulges in the bird's neck. The occasional starfish they brought up seemed to be less attractive to them; they rarely did more than nibble at it or shake it and eat one or two arms before they abandoned it.

Some Eiders had discovered an easier way of earning a living. In Seahouses Harbour a few could always be seen swimming round the fishermen's cobles. They had lost their fear of Man entirely and greedily swallowed offal thrown overboard. And when that other fisherman, the Grey Seal, came up with a large Lumpsucker—a common sight on the Farnes—some Eiders often joined the gulls and feasted on the remains discarded by the Seal.

Early in spring the Eiders began to concentrate near the islands where they were going to breed. Of all the Farne Islands ours was the most popular. Already in February we saw dozens of pairs in the Kettle. Before sunrise they came drifting in on the strong tide. I shall never forget these glorious mornings. On calm days the water of the Kettle glowed in the most wonderful colours, partly caused by the reflection of the purple eastern sky, partly by the varying shades of brown and green in which the weed patches and the clean sand of the bottom showed through.

Most of the birds swam in pairs and although we were rarely able to distinguish and recognize individuals at that time, we often followed individual birds for long periods through our binoculars and found that, however much they might get mixed up with other pairs, they always returned to their mates. There can be little doubt that by this time these birds were paired and that each bird could distinguish its own mate from all the other Eiders. There was by now almost continuous display, fighting and courting—never a dull moment in the Kettle!

The males were not less quarrelsome than other birds in spring, but Eiders' fights were rarely impressive. Usually we saw no more than a short scuffle; one male would suddenly make a dash at another who happened to come too near and the other immediately threshed away. Sometimes the attacker got hold of his adversary, but then he rarely knew how to do anything more heroic than just hold on, while the other struggled, rather meekly, to get free. When the short skirmish was over, the beaten bird would shake itself or stand up and flap its wings, and that was all. But on those rare occasions when the attacked bird really stood up and fought back we saw magnificent battles. Grasping the most readily available part of the opponent's body—bill, neck, wing, tail—each bird pushed, pulled and twisted its rival with furious force and in the meantime thrashed him with powerful wing-beats that could be heard from a great distance. Locked together the two champions would roll over and over and it was rarely possible to disentangle the whirling mass of wings, bodies and spray.

All through March and April the numbers of Eiders in the Kettle increased. Yet until well in May they seldom came on land, or at any rate not on our island. About noon small groups used to gather on the low shingle banks of Knox's Reef, where they were surrounded by water. Here they dozed, roosted and preened their plumage.

But early one May morning, as we woke up at dawn to the ringing of the alarm clock, we could hear the cooing of the drakes and the 'kokokokok' of the ducks much nearer than before. Looking out of the window we saw that several pairs had landed. The invasion of our island had begun.

This annual invasion was always fascinating to watch. Shortly after dawn pairs would begin to approach the land, at first no more than a dozen of them, but soon their numbers ran into the hundreds. Their approach was cautious and hesitating. When there was a swell, they

showed admirable skill in avoiding being bumped on the
rocks. Just behind the breakers they would ride for a
considerable time, then all at once make up their minds
and very quickly let themselves be carried on the crest
of a wave, landing and running a dozen steps before the
next wave could overtake them. Then they paused,
flapped their wings, preened a little, and began their
slow, laborious pilgrimage uphill, waddling on their
clumsy legs, hopping over small cracks, skidding and
tumbling over the slippery seaweed, but advancing
steadily. From now on we could really call the ducks
our neighbours, for watching them from our room in the
tower we saw to our surprise that most of the invading
ducks chose to proceed to the area immediately surrounding the tower. In whatever direction we looked, we
always saw Eiders right near our house. Some pairs even
walked into the courtyard in front of our only door.

This created a problem. For our work we had to go
out at dawn. By getting up early enough we could slip
out before the Eiders came. But from then until the moment they decided to return to the sea—usually round
9 or 10—we could not move freely near the tower. Whoever was outside had to stay out; he who lingered in
bed too long had to stay in, or else the whole crowd
would fly off in panic. Luckily the ducks soon became so
tame that we need not bother too much and in the three
seasons we spent on the island neighbourly relations were
as good as one could wish.

It was highly interesting to watch this cautious approach to the breeding grounds. Like so many other sea
birds—for instance gulls and terns—the Eiders were torn
between two urges: the urge towards the breeding
grounds and that towards the safe, open water. On land
they are pretty vulnerable; yet they have to nest on land.
Their start is extremely cautious and wary, but once they

have been on land for some time unmolested, they become very tame.

Once the invasion had started I often spent the early mornings in the tower, watching the goings-on from the window. When the alarm clock went off, I had only to jump out of bed, wrap myself in my duffelcoat and walk five steps to the window. Here, behind a cardboard screen, I could watch unobserved, in the meantime lighting the Primus for the indispensable mug of coffee, and have a snack as well.

From this citadel we had a wonderful view of the ducks and could study their displays. The drakes continually produced their sonorous, dove-like cooing calls. They uttered only three of them and it must be admitted that their conversation was rather monotonous. Dr Frank McKinney, who made a special study of the Eider ducks on the Inner Farne, found that on the whole this cooing was very much mixed with hostility between the drakes. The motivation of the birds during their courtship performances could not be understood so easily as in the gulls, but it was clear that Eiders, too, are in a state of great tension when they court females and that aggressiveness and fear were often mixing with the mating urge.

Many species of duck are known to have a peculiar habit, in some ways unique among birds: the females have a special display by which they incite their mates to attack strange drakes. The Eider is no exception to this rule. This is a most curious affair, so curious indeed that many animal psychologists, particularly those working in laboratories where rats and pigeons rather than ducks are studied, simply do not believe that this kind of thing happens at all.

When a pair is approached by a strange drake, whether or not accompanied by his own wife, the female gets annoyed. She points her beak towards the stranger, stretching her neck in his direction, and then, withdrawing her neck, makes a few upward movements with her head. Then she points again at the stranger, followed by the 'chin-lifting', and so on. All the time she utters a querulous, harsh 'kokokokok, kokokokokokok'. Her husband—and this is perhaps the most curious aspect of the whole business—cannot resist this. He may have been dozing or preening and he may not at once respond to her 'inciting', but when she persists (as she usually does) he gets up and attacks the stranger. When there is more than one stranger in the vicinity (and there may be more than a dozen), he invariably attacks the one that his wife pointed out! Several times I saw how a female switched her attentions from one male to another and it was pathetic to see how her mate, just (almost reluctantly) preparing to attack the male she had been pointing out first, switched to the new object of her displeasure. The females at this time of the year are not a very pleasant lot.

To the zoologists the inciting movements offer several puzzles. One is that of their origin. In many species this is not easily understood, but the behaviour of the Eiders gives us some indications. The first part of the movement—that of pointing to the strange male—is an expression of aggressiveness. It is nothing but an incipient attack. The female points to the male in exactly the same way as the male (or for that matter she herself) does when pecking at a rival. If the female is daring enough she may actually proceed to bite the male she is pointing at. The second part of the movement—the repeated chin-lifting—expresses an internal conflict in which fear is one (and the dominant) component and a tendency-to-stay another. This is obvious when one compares the various situations in which chin-lifting occurs. We saw it done by birds which were fishing at the foot of the cliff when they suddenly discovered us; they were just a little afraid, yet reluctant to leave the feeding grounds. We saw it also in females on the nest when approached by us and regularly in mothers with chicks when we came too near. When we were watching and photographing from a hide Eiders that were coming ashore to drink in freshwater pools we saw them do the chin-lifting every time they hesitated on their way up to a pool near the hide.

In all these situations the birds showed slight fear, yet there was always something that compelled them to stay. But the reason why they wanted to stay was different in each case. I believe that an inciting female is afraid of the strange male, but wants at the same time to attack him, and yet to stay near her own mate. It further looked as if the chin-lifting made it difficult for a male to attack her—as if it functioned as appeasement. It seems, therefore, as if we have to interpret the inciting movement as a sequence of incipient attack and an appeasement movement. The incipient attack makes the male join in and even carry out, on his own, the real attack; she protects herself against getting involved in the actual fight by showing the appeasement movement. As seems to be the case in so many other signal movements in animals, the chin-lifting of the female seems to be the mere outcome of a relatively simple emotional state of the bird. Somehow this movement has acquired an extremely useful function as a signal, probably without the individual birds knowing it at all.

Soon after first coming on land, the ducks began to look for nest sites. This, and the subsequent building of nests, offered us a good opportunity for observing the gradual development of a seasonal behaviour pattern. At first, females would sit down here or there, making a few incomplete building movements in a half-hearted, desultory way, then stand up again and walk on, sit down ten yards further on, and again make some feeble scraping movements with the legs; or they would look intently at the ground for a moment, perhaps even take up some straws, but then drop them again, then abandoning this site too, and so on. The drakes accompanied their wives and even made a few incomplete building movements themselves, but unlike the females they never went beyond these tentative beginnings. The females grew more efficient from day to day and finally scraped out a beautiful round cup, lined it with straws and whatever other material was lying round the cup, and finally laid the first egg—an event which we witnessed several times.

Not until the clutch was almost complete did the ducks begin to brood continuously. Until then they covered

Fig. 22. Eider Drake displaying.

the eggs with loose nest material every time they left the nest. This was a very useful ruse; it helped to protect the eggs from robbing by the ever-present Herring Gulls and Lesser Blackbacks. It was not entirely successful, for some gulls managed to uncover Eiders' nests; dozens of broken egg shells told of these gulls' success. But most clutches escaped detection. Luckily the gulls were pretty stupid even though they could uncover some nests; we watched one yearling Herring Gull which knew in his dim way that eggs were hidden under plants, but he never knew where to look for the nests and often went about, walking aimlessly all over the place picking up and throwing aside any loose material he met—a curious mixture of cleverness and stupidity.

Soon one duck after the other began to sit. Already before this new period began, the birds had, in a relatively short time, become surprisingly tame. Several pairs had walked into our yard and when we met them there, they scarcely moved out of the way. They just looked at us a little suspiciously (scrutinizing us so intently that we felt inclined to apologize for disturbing) and that was all. Both sides considered the other a nuisance, but both accepted the inevitability, like families sharing the same kitchen.

This was also the time when the famous eider down began to appear in the nests. Watching from our citadel we had a fine opportunity to see how that came about. We counted now about fifty ducks sitting within a distance of 40 yards from the window. Now and then a duck would stand up and persistently and vigorously preen its breast or belly, wriggling with the bill tip wedged deep into the plumage. When it withdrew its bill, one or a few downy feathers could be seen sticking to its tip. With a slow shaking movement, often touching the nest with the bill, the bird got rid of the down and then started down-preening again. In this laborious way, bit by bit, the lovely down lining of the nest was built up.

The sitting ducks were so devoted to their task or, rather, so keen on sitting, that they never took time off to eat. This may sound incredible, but it is the truth: after having produced their huge eggs—in itself a strain on the body's resources—they just starved for about a month. They did leave the nest every second or third day for 10 to 15 minutes at a time, but only went to drink.

The end of May and the first half of June were a marvellous time on the Inner Farne. The island, usually a pretty bare place, was by now adorned with dense carpets of lovely flowers, mainly Thrift, Sea Campion and Silver Weed. All the breeding seabirds had arrived at this time. The Kittiwakes had built their nests on the South Cliff, the Shags were feeding their young on their huge nests; colonies of those gaudy clowns, the Puffins, had established themselves in the turf on the southern and western ends of the cliff. The sand beach in St Cuthbert's Cove and the belt of low rock along the Kettle were covered with the nests of Arctic Terns, whose shrill voices filled the air. On the top of the island there were Common Terns mixed with the Arctic and even a few Roseate Terns. One year a fair-sized colony of Sandwich Terns sprang up. There was little room on the island for anybody but birds and if we had by now failed to realize that we were intruders in this birds' world, the fierce Arctic Terns reminded us of it by their incessant vicious attacks.

However, the decline was to begin soon. The Eiders were about to leave. The drakes had long gone. The first cracks began to appear in the eggs. The females had grown very thin.

The hatching of each egg took a considerable time and even after the chicks had appeared the mother covered them for hours. The ducklings, who at first lay quietly in the nest, soon became restive. The first sign of this was the fact that the mother had to stand up higher and higher, tolerantly allowing her lively offspring to knock her about. Before long we would get a first glimpse of something dark and fluffy under her tail or wings. Soon afterwards a tiny head would appear and two alert little eyes were looking out into the world. Few more touching scenes can be imagined; just as with the adult Eiders, one has to see these perfect babies to believe that they are true.

In a few hours the chicks showed a rapid development of behaviour. They started by preening themselves thoroughly, until all the débris of the sheaths that had surrounded their downy feathers had been shed. Then they began to half walk, half crawl about with small and very uncertain steps. Soon they walked pretty confidently and began to nibble at the flowers near the nest. And then the great moment came: all at once the mother stood up, walked a few steps away from the nest and sat down again. The ducklings got up and approached her. She might cover them again, but soon she would make a few more steps, utter a couple of 'koks' and the chicks would follow again. And so, in easy stages, the family began to move away. Often they walked down to one of the tide pools, where they made another stop. Here the chicks took to the water at once, swimming about and diving energetically in a flurry of spray. And then it was a matter of a few hours before the family walked off to the sea and swam off. That would be the last we saw of them—until the following year.

The strain that four weeks' uninterrupted incubation put on the ducks was revealed several times when females deserted their broods in sight of harbour. We saw several that just gave up, in spite of the fact that the eggs were already pipped. One such case I remember vividly. A duck nesting in the yard, who had stuck to her task bravely for weeks, walked off one afternoon at three. We happened to see her go and were struck by her poor condition. She was very thin and her steps were faltering. Every ten yards or so she sat down and rested and it took her twenty minutes to cover fifty yards. When we looked at the eggs we saw that they were about to hatch; there were large holes in them and we could hear the ducklings squeak inside. By five o'clock the mother had not returned and the eggs were cold.

It was then that my friend Olaf Paris, who was staying with us in the tower, decided to take over the mother's job. As it was quite cold weather, we had a fire in our room. Nest and eggs were taken indoors and placed, in a cardboard box, in front of the fire. A hotwater bottle was added. The chicks revived, but they were not yet ready to hatch. So, at nightfall, Olaf took the whole box with him into his sleeping bag—appropriately made of eider down. He spent an uncomfortable night playing the mother duck. Whenever I woke up I heard him talk to his charges: 'Kokokokokok! kokokokokok'. At five in the morning there was general rejoicing at the happy event in the Study Centre. Olaf, beaming and kokkok-kokking happily, produced first the egg shells and then five charming ducklings. They were as healthy and contented as could be.

Next we looked out of the window and inspected our Eider-duckery. And sure enough, there was one mother in the same blessed circumstances as Olaf, squatting contentedly with four ducklings under her. Frank and Olaf went down with Olaf's brood and very slowly approached the duck. Cautiously, without flushing her, they put the ducklings under her; and so the duck was saddled with a brood of nine. All went well. The mother did not object and neither was there any friction among the ducklings. A few hours later the whole family marched off to the sea.

The trip from the nest to the sea was always fascinating to watch. A family rarely made it alone; they were usually accompanied by 'aunts'. This was Frank McKinney's appropriate name for stray females who were at this time hanging round the island, probably ducks who by some mishap had lost their own brood. As soon as such birds discovered a mother with chicks, they were irresistibly attracted to them and soon a group of three, four, or even more aunts would collect round a family, often even before the mother had shown any inclination to walk off.

The innocent ducklings were just as trusting towards the aunts as to their own mother. Once they began to wander around, they often waddled towards an aunt. And then the aunt showed a curious response: when the chick came too near, she would back away. Occasionally an aunt pecked at such a chick and that was why it was necessary that the mother should resent the too-near presence of any aunt. She made a dash at them now and then, keeping them a couple of feet away. This curious mixture, in the aunts, of parental interest, fear and aggressiveness was very revealing; I am inclined to believe that in the normal mother, parental love, although absolutely dominant, may also be mixed with traces of fear and aggressiveness too.

When the family walked off, the aunts went with it. Often several families marched away together, surrounded by the whole escort of interested aunts. After the usual stop in the tidal pool, the journey was continued and now it reached its spectacular climax. The mothers led. Calling continuously, they waddled down to the sea, jumping over clefts in the rocks, fluttering down precipitous stretches. The ducklings followed as fast as they could. Most mothers went down the gently sloping rocks bordering the Kettle, but several went to quite steep parts of the cliff. The first time we saw such a mother heading for the cliff we were prepared to see disaster and were sorry for both mother and chicks. But to our amazement the chicks, with complete and touching confidence, just jumped after the mother. They sailed through the air, tumbling round and round on the way down, 10 or 20 feet, and fell with a bump on the hard rock underneath. But they got to their feet at once and went on as if nothing had happened! Another jump, and one more, and they landed on the sea. Within a few seconds the ducklings had all gathered round the mother and at great speed they set off to the west, to Budle Bay, where most Farne Island Eider families gather. There they fed in the shallow water, well out of reach of predators.

This dramatic departure, which we witnessed a score of times, did not always go smoothly. Sometimes ducklings got stuck in crevices. Their desperate squeaking stopped the mother for a while. She waited and might even come back to the unfortunate chick. But she would not offer any real help and if the situation was hopeless I think she went off without the chick, leaving it to die, either of exposure or, if it was lucky, under the merciless bill of a Herring Gull.

When we discovered such chicks in time we could sometimes save them by throwing them into the sea as far as we could, trying to drop them as near as possible to their own mother. This method looked rather brutal, but the chicks always landed safely and usually the mother, once she saw and heard it, waited for it.

Esther and Mike Cullen once saw an Eider mother with five chicks preparing for a jump down the high cliff. They saw all five chicks jump, then, for a short while, they disappeared from sight, and when the family reappeared on the water there were only four chicks and the mother was hesitating and looking back at the cliff. Guessing what had happened, Mike and Esther hurried down and found the fifth chick wedged in a crack from which it could not free itself. Upon their approach the mother started to swim off with her four and by the time the chick was freed she was some fifty yards away. A Herring Gull had discovered the chick at the same time as the Cullens and, intent on a tasty bite, it circled over the scene. There was no time to lose and since there was no other Eider family in sight, Mike hurled the chick out to sea. The poor thing fell on the water with as much of a thud as such a small body could. At once the Herring Gull went after it and since the chick seemed hurt and began to swim in cricles, its prospects did not seem too bright. But the Cullens decided to put up a fight. Every time the gull swooped down, they jumped into the air, madly waving their arms and yelling a fierce war cry. Each time the gull was scared off for a little while, but it returned persistently. Luckily just in time the chick recovered and started to swim away from the island, calling its distress call all the time. This saved it; the mother heard it and turned back, the four other ducklings keeping close to her. She reached the chick and from now on she took charge. The gull, never keen on meeting a furious Eider mother, gave up and the battle was won.

Gulls were always trying to get at Eider chicks. But the Eider mothers—and also some of the aunts, bless them—defended the chicks vigorously. Upon the approach of a gull the mother called the alarm, a long drawn, grating 'rooo!' and then the chicks hurried towards her and crowded round her in a dense little cluster. Usually this was enough to make the gulls give up—at least for the moment, for the gulls, though surprisingly timid when meeting with resistance, were patient and persistent. They alighted nearby and awaited their chance. And when that chance came in the form of a stray chick which had left the safe vicinity of adult ducks, a gull would suddenly swoop down and, before the growling ducks could prevent it, flew off with the pitiable little thing struggling in its bill.

But such is life, at least for Eider ducks. Nevertheless, many ducklings survived and on our occasional visits to the mainland coast we met numerous flourishing families.

And so, with the hatching of the chicks, our neighbours the Eiders left us. At last we could walk freely over the island again without feeling the silent reproach in the eyes of the ducks crouching over their broods. For that was somehow more difficult to ignore than the vicious attacks of the Arctic terns.

CHAPTER FIFTEEN

INSECTS AND FLOWERS

The wonderful symbiosis, or mutually beneficial relationship, between flowers and insects, so well known at present, can be regarded as the discovery of Christian Konrad Sprengel, a German schoolmaster who, in 1793, published (to quote Darwin) 'a remarkable book with a remarkable title': *Das entdeckte Geheimniss der Natur im Bau und in der Befruchtung der Blumen* (Nature's Secret Revealed: The Structure and Fertilization of Flowers). Sprengel's book attracted little attention until it was rediscovered by Darwin.

His work did not lead to experimental studies until the famous German zoologist, Karl von Frisch, started his long series of researches into the behaviour of the Honey Bee. Von Frisch's paper on the colour sense and form vision of the Honey Bee, and the controversy with Von Hess (who thought that Honey Bees must be colour blind) initiated intensive and extensive work on the relationships between flowers and insects. Von Frisch himself showed that Honey Bees respond to certain colours, to a lesser extent to the shapes, and also to the scents of the flowers they pollinate. Often bees are attracted first by the flowers' colours and, just before alighting, check its scent; if it is 'correct' they alight, if they perceive no scent, or the wrong scent, they hesitate or even abandon the flower altogether.

The Austrian botanist, F. Knoll, expanded this work by studying insects other than Honey Bees and by studying the adaptations of different types of flowers. Knoll's work made it clear that there are many types of flowers, each adapted to a different kind of pollinator. His studies of the Bee Fly, *Bombylius*, in Dalmatia showed that these insects respond to flowers in roughly the same way as Honey Bees. When they visit Grape Hyacinths, for instance, they respond above all to the blue colour of their flowers.

He found a quite different arrangement in the inflorescences of our common Arum lilies, *Arum maculatum*. These flowers attract small insects of various kinds, mostly Diptera, by their scent and trap them. They have a beautifully adaptive combination of slippery surfaces and one-way 'fences' of hairs and once the pollinators are inside, they are kept there, and only after they have covered themselves with the pollen are they allowed to get out. And, being pollen-covered, they at once proceed to get trapped in another inflorescence and, since in these plants the female flowers mature first and the scent is only produced during this stage, they invariably land in an inflorescence ready to be pollinated. After they have done their duty, they are detained for a while until the male flowers open, when they become covered with a fresh load of pollen.

Other studies by Knoll concerned hawk moths. He found that the typical hawk moths, flying at dusk, responded, even when it was almost dark, to the colour of various flowers. The Hummingbird Hawk Moth, which flies in bright daylight, is even more obviously visual in its responses. Knoll's observations on this species are rightly famous. He demonstrated, for instance, that this moth responds very accurately to so-called honey guides. The Common Toadflax, a flower with concealed nectar produced inside the long spur, has a bright orange patch on the lower lip, near the slit-shaped entrance to the spur. Knoll pressed Toadflax flowers behind a glass screen and observed the exact spot where the moth, hovering in front of this dummy, touched the glass with its unrolled tongue. Each time the moth touched the glass with its tongue, it left a little smear of the sticky sugar-solution on which it had previously been feeding. With *Lycopodium*-spores Knoll made these spots visible and so could produce a graphic record of the moth's 'hits'. He cut the orange spot out and fastened it in front of various parts of the flower and even then the moth always aimed at the orange spot.

Knoll was also much interested in the function of brightly coloured leaves, which are not really part of the flowers themselves. Thus *Salvia horminum*, a Mediterranean Sage which is popular in British gardens, has relatively small and inconspicuous flowers, but bright purple or violet leaves at the top of each shoot. When Honey Bees begin to visit these plants early in the season, they invariably fly towards these bright top leaves and, not finding any nectar there, slowly descend towards the real flowers. Experienced bees respond directly to the flowers.

These studies made it clear that the relations between flowers and pollinators are extremely elaborate and varied and although more and more biologists are working on these subjects, and new aspects are being discovered almost every year, the subject is by no means exhausted. Our knowledge of tropical flowers, for instance, is still very scanty, in spite of the many interesting facts reported, for instance, by Van de Pijl, who gave highly pertinent descriptions of the many peculiarities that must be considered adaptations in flowers pollinated by bats.

In the course of our own studies of insect behaviour we could scarcely avoid doing something on this subject as well. In Chapter Nine I mentioned that Graylings and other butterflies responded to the scent of flowers in a way slightly different from bees. They would not respond to coloured papers unless a flower scent was sprayed into the air. As soon as this was done, however, they began to fly straight at the coloured papers and even alighted on them, although the papers themselves did not emit any scent. Scent, therefore, only alerted them and instead of responding, as bees do, to the colour first and then making a close inspection of the scent, they did not react to colour until they perceived the odour.

Philanthus behaved in a different way again. It, too, showed little interest in paper models of *Erica*-flowers even when confined with them in small cages. When Van Beusekom tried to work the same trick on them that had worked so well with the Grayling, and hung a bag with scented *Erica* to windward of the cage, the wasps at once began to fly against the wind towards the bag and bumped into the gauze of the cage. This species, therefore, was also alerted by scent, but unlike the Graylings it flew towards the source of the scent.

In other years we had occasion to make a few observations on the behaviour of Pine Hawk Moths, which were very common in the extensive pine plantations where we camped. After reading Knoll's reports about the responses of other hawk moths to the visual properties of flowers, I always felt that this could not be the whole story, because many of the flowers visited by moths flying at dusk start to emit a very strong scent at that time of day. From the extent of the scent clouds hanging round flowering Honeysuckle bushes, for instance, I thought it very probable that moths perceive the scent when they are still so far from the flowers that they cannot possibly see them. This could easily be tested by setting up Honeysuckle flowers at dusk, concealed in a wooden container which screened the flowers from sight but allowed the scent to disperse freely through a number of slits. Many Pine Hawk Moths responded to this set-up. All approached it against the wind and in a zigzag course; some from a distance of 10 yards, crashing through the crowns of the young Pine trees surrounding it, and all finally found their way in through the slits.

It was not until later that one of my co-workers did more systematic work on the behaviour of insects towards flowers. In 1952 Aubrey Manning decided to study the behaviour of bumblebees. He was particularly interested in three problems. First, he felt that more should be done about honey guides. Knoll's observation on Toadflax had remained rather isolated; there were so many other types of alleged honey guides to be found in various flowers, such as the radially arranged lines or the striking patterns of dots round a flower's 'entrance'. Before the war, the German, Kugler, had done a great deal of work on bumblebees, but his conclusions on the part played by honey guides did not seem correct; briefly, he claimed to have shown that honey guides do not actually 'guide' the insects to the centre of the flower, but that the only thing they did was to make the flower as a whole more attractive thus and to release more, not better aimed, visits.

Second, there were suggestions in technical papers that the total reaction chain shown by an insect when it visits a flower is longer and more complicated than was evident from von Frisch's work on the Honey Bee. I had myself observed that bumblebees, visiting flowering plants of Houndstongue (*Cynoglossum*), often made mistakes; they flew to plants such as Ragwort or various thistles that were roughly similar to Houndstongue in general appearance; yet these were not themselves flowering at that moment and the bees did not fly to the places where these plants would bear their flowers (at the top) but to the places in the axils of the leaves where

Cynoglossum usually has its flowers. It looked as if these bumblebees responded to the general appearance of the plant first and did not see the flowers until they were much nearer to them. I had also noticed that some bumblebees seemed to know by experience where individual plants were growing. Some bees that we could recognize by colour dots which we had given them visited a certain group of plants at regular intervals; they flew from one plant to the next in a fixed sequence and, when I pulled a plant out, the bees kept returning to the site where it had been, searching at the spot for a considerable time before abandoning the attempt. All this suggested that the behaviour of bumblebees was much more complicated than we had hitherto suspected.

Manning's third problem concerned the possibility of a 'language' of bumblebees comparable to that analysed so admirably by von Frisch in the case of the Honey Bee.

Manning spent three seasons studying bumblebees and succeeded in discovering some extremely interesting new facts. His search for a 'language' gave an entirely negative result: nothing of the kind seems to exist in bumblebees. He found no evidence of 'scout' bumblebees alerting others when they had discovered an abundant new source of food. In this respect bumblebees are absolute individualists. Manning thinks that the absence of 'language' has to do with the small size of bumblebee colonies. It may be that a signalling system is useful only in large communities where the foragers concentrate on mass-crops.

More positive information was gained about the effect of honey guides. Manning did his observations on Oxford University's estate near the small village of Wytham, a wooded area of approximately 1,000 acres, with patches of arable land and some young plantations. Various species of bumblebees (*Bombus agrorum, B. pratorum, B. lapidarius, B. terrestris* and *B. lucorum*) were common in this area. In order to have numbers of bees available for tests whenever he wanted, Manning dug out a number of nests, accommodated them in nest boxes (designed by my friend Dr Jan Wilcke of Bennekom, Holland) and put these up on shelves in a little hut on the estate. Each nest box was connected by a plastic tube with the outside world and, in order to facilitate correct homing under these crowded conditions, the exits of the different tubes were painted in different bright colours. These bumblebees resumed their normal routine soon after the transfer and, while foraging, acquired a thorough knowledge of the surrounding area. Whenever Manning wanted to subject them to tests with models of flowers, he erected a flying cage of $8 \times 4 \times 4$ ft. in front of the hut. Some bees did not adapt themselves to this interference with their normal flights and kept buzzing and bumping against the gauze; but others soon responded to the flower models inside the cage (which were often supplied with sugared water) and established a regular shuttle service between them and their nests.

Manning started by feeding his bees sugar-water from a little dish put on the centre of a rectangle of bright blue paper, measuring 15×10 cm. Once the bees visited this regularly, he took the rectangle away and offered, under a clean glass pane, three figures cut of identical blue paper, one in the shape of a circular disc, one in the shape of a star, and one of a shape roughly like a huge Primrose (Fig. 23). No sugar-water was offered. He now

Fig. 23. Models for attracting bees.

observed how well the bees could find the centres of these models. They approached every model, but (presumably because no food was offered) did not alight. They hovered about one inch above them and now and then made a quick dip down, almost touching the paper, which Manning interpreted as incipient landings. These dips were usually aimed at the edge of the models, relatively few being directed at the centre. The circle received 66 dips on the edge and 5 on the centre; for the star these figures were 22 and 1, for the 'Primrose' 36 and 2.

These responses, to large, plain model flowers without anything like a honey guide revealed an interesting tendency on the part of the bumblebees: they were obviously attracted mainly to the edge, to the line of contrast between the blue and the colour of the background. These tests, therefore, were an excellent starting point for an investigation into the possible guiding effect of honey guides: if figures similar to honey guides were provided, would the bees be drawn towards the centre? Before trying this, Manning did a slightly different test: he fed his bees on the star, providing sugar-water on the centre. Then he presented the same star but without sugar-water. Now he saw 180 dips aimed at the edge and 43 at the centre. So even after being trained to find food in the centre on a flower model of which, one would think, it should not be difficult to find the centre, the great majority of bees still dipped at the edge.

Manning soon found that this 'edge effect' was noticeable in bumblebees foraging on wild flowers as well. When they visited small flowers it was not easily detected, but as soon as a bumblebee alighted on a large flower, such as a thistle, it usually alighted on the edge.

This result made it possible to study the effect of honey guides. A new training model was provided—a star with a bright yellow line on each 'petal'. Six of these lines radiated towards the centre, where sugar-water was again provided. After training, the same model was offered without sugar-water. And now Manning counted 103 dips at the edge and 172 in the centre! This was a clear indication of the guiding effect of such lines. Just to make sure that this had been the effect of the radial pattern, he now presented a plain star and got 103 dips on the edge and only 15 in the centre. When he did a similar test with a star showing a circular yellow dot in the centre, this dot also acted as a guide. (See Fig. 24.)

Fig. 24. Models for attracting bees.

In another experiment Manning fed the bees on a model without a honey guide and then presented one with it. In this test, too, there were more dips on the centre. This suggested that the bees responded to the honey guides without having learnt them. However, they might have become acquainted with all kinds of natural honey guides during their unimpeded foraging flights in the country.

Contrary to Kugler's conclusion, therefore, Manning found clear evidence of a true guiding effect.

Next the question arose as to the maximum distance from which the bees could see the honey guides. Kugler seemed to assume that they responded from a long distance and were more attracted to flowers with than to flowers without a honey guide. Manning thought this improbable because the visual acuity of bees is probably too poor for this. He now fed the bees on two circular discs of 4 in. diameter, one plain, the other with a honey guide. Then he offered both models without sugar-water and watched the course taken by the bees. At a distance of about 2 ft. it was clear at which of the two models a bee was aiming.

The responses to the plain model were just as numerous as those to the honey-guided one. And it is quite probable that at this distance the bees could not really see the difference between the two models. But once they had approached a model, it was obvious that they distinguished between them: they dipped much more often at the honey-guided model. On both models the first dip was usually directed at the edge, but the subsequent dips were aimed much more predominantly towards the centre in the honey-guided model than in the plain one. The figures in one of these tests were: Model with honey guide: 101 first responses to edge, 2 to centre; 60 subsequent responses to edge, 121 to centre. Plain model: 142 first responses to edge, none to centre; 86 subsequent dips to edge, 26 to centre.

The following year this test was repeated with bees which had seen only plain circular models. The honey-guided model now received 128 first dips to edge, 8 to centre; subsequent dips: 45 to edge, 232 to centre. Plain model: 119 first dips to edge, 3 to centre; 124 subsequent dips to edge, 58 to centre.

Bees, therefore, always start to fly towards a flower's edge; if there is a honey guide they soon find their way to the centre; if there is none this is much more difficult. Manning saw a beautiful demonstration of this in bees visiting large *Magnolia* flowers. They flew again and again to the edges and the tips of the petals. Most of them found the flower's centre after some searching, but several never achieved this and finally gave up!

The same tests allowed Manning to compare the total number of dips elicited by plain models with those towards honey-guided models and so to test Kugler's suggestion that honey guides increase the attractiveness of a flower. In tests with honey guides the average number of dips per visit counting first as well as subsequent dips was 2.75 in one series and 3.1 in another. In the same tests the plain model received on the average 1.8 and 2.5 dips. Honey guides, therefore, do make a flower more attractive, but the effect appears only after the bee has been attracted to the flower from afar. The honey guides made the bees more eager to alight once they had arrived.

In all these tests the artificial honey guide consisted of yellow lines on a blue background. In many flowers, however, the alleged honey guide is not so strikingly different in colour, but is of a more intense hue of the flower's general colour. Manning's method enabled him to test the effect of such patterns as well. He fed his bees again on a plain blue model and in the test offered this model together with a similar model which, however, had a deeper coloured centre. The plain model elicited: 80 first dips to edge, none to centre; subsequent dips: 107 to edge, 32 to centre. The other model got 59 first dips to edge, none to centre; 65 subsequent dips to edge, 66 to centre. The deeper hue, therefore, was a more efficient honey guide.

Fig. 25. Models for attracting bees.

In this way Manning penetrated further and further into this fascinating relationship between flowers and their pollinators. I need not go into more details; the general procedure is clear and essentially very simple.

Manning himself would be the last to claim that he had exhausted the subject. For instance, he points out that one would now expect that large flowers would be more in need of honey guides than small ones. Yet many of the most intricate honey guides are found on small flowers such as Forget-me-nots, Speedwell, and Eyebright. These flowers are visited predominantly by small flies; how would these respond to the honey guides?

Further, his work concerns visual honey guides only. Through the work of Lex, one of von Frisch's pupils, however, we know that many flowers have 'chemical honey guides' as well. She found that many flowers emit either a more intensive scent, or even a different scent from the radial veins of the petals, or from small dots round the centre, than from the rest of the petals, and she showed that Honey Bees respond to such chemical honey guides.

A couple of hundred yards from Manning's 'bombarium' there was a lovely open part of Wytham wood where scattered plants of Houndstongue attracted quite a number of bumblebees during May, June and July. Here Manning spent many days investigating their behaviour while foraging on these plants. Houndstongue does not flower until its second year; among the large, coarse leaves the small flowers, though numerous, are not very conspicuous. They produce abundant nectar and that is probably why many bumblebees specialize on them entirely. This fact offered good opportunities for a study of their learning abilities.

In the beginning of the season some time elapsed before the bees discovered the flowers. In 1954, for instance, the first flowers in the group which Manning kept under observation opened on May 14th; on the 20th several of the flowers produced abundant nectar, yet the first bumblebees did not visit them until the 26th. Manning was sure that he did not miss early visits, for those

Fig. 26. Houndstongue.

early flowers never set any seed at all. These first bumblebees had not yet specialized on Houndstongue, but carried pollen of many different flowers. They never reacted
to the plants from any great distance; they never visited
plants without flowers; and they flew only towards the
flowers when they happened to come within a few inches
of them. Already at their second or third visit, however,
these bees showed the first signs of getting acquainted
with the species.

By giving a number of the bees small colour dots on
the thorax, Manning marked them as individuals and
so could follow their behaviour through several days.
The first signs of learning were occasional visits to plants
which, though of entirely different species, had a general shape much like Houndstongue plants: Ragwort,
Thistles, and St John's Wort. Unlike 'naive' bees they did
not merely respond to the flowers themselves, but flew
directly towards such plants from several yards away
and then searched at exactly those parts of the plant
where a Houndstongue had its flowers.

Manning noticed soon that the regular customers developed a fixed routine. In a group of Houndstongue plants, with distances between the individual plants ranging from one to a couple of feet, each bee visited the plants in a more or less fixed order. Once they were fully laden they left for home, but returned at regular intervals. Manning now pulled out one of the plants of the group. When one of his regular bees returned, it did exactly what I had seen some bumblebees do in Holland many years ago—it flew to where the plant had been and circled round the spot for a long time, as if searching for it. It might be a minute before such a bee gave up its attempts and flew to the next plant. One of these bees which returned every half hour, gave the following record: the first time it searched for 40 seconds, when next it came back it circled for 7 seconds, next time 1 second, at the fourth visit it just flew to the empty site but flew on at once, and not until the fifth time did it skip the site altogether. Manning found that this site-conditioning was evident only where the plants stood rather far apart; in the dense centre of a group the bees seemed to rely on being able to see the flowers directly; the sequence in which they visited the individual plants was also less regular there. It seemed, therefore, that they learned the site of groups of plants and of individual plants only where each one stood well apart of the rest.

To make sure that the bees were not reacting to a trace of scent which might remain hanging round after a plant had been pulled out, Manning repeated the test with potted plants which he could move about at will without damaging them. When he got the same results, he checked the possible influence of scent in various other ways. For instance, the bees never visited the yearling Houndstongue, which were nothing but flat 'rosettes', but which gave off the typical Houndstongue-scent, though admittedly no flower-scent. Neither were the flights towards intact plants ever dependent on the direction of the wind as must be expected when the bees found them by their odour.

This set-up allowed Manning to see how bumblebees discovered new plants. Some of them were quite satisfied with the few plants which they visited regularly; they came straight to the central group, drank their fill and made a 'bee-line' for the nest. Others, however, left the central group now and then and flew off in an irregular, meandering course; these were the bees which approached any plant looking remotely like a Houndstongue. Manning now put up a few of his potted plants a couple of yards from the central group. These plants provided abundant nectar, as they had not been visited continually by foragers. Such plants were soon discovered by the wanderers. Once there, they naturally stayed a long time. When satiated, these bees showed a characteristic behaviour before leaving. They flew up, made a short loop through the air, flew back towards the plant, made again as if going to fly off, but returned once more, and thus made a series of ever wider loops before finally leaving. In short, they made what looked exactly like a 'locality study', such as we had seen so often in *Philanthus*. Manning showed that this was its function by taking such a plant away immediately after a bee had made such a locality study; when next the bee returned, it showed that this one locality study had been enough to imprint the site of the new plant on its memory—it searched round the spot where it had been during its one previous visit.

And there the story would have ended if Manning, as a true naturalist, had not kept wondering about the curious difference in behaviour between bees foraging on the central group and those visiting isolated plants. In the central group, as we have seen, the bees did not show such clear signs of attachment to individual plants. Was this perhaps because they relied, so to speak, on having the other plants within easy reach, so that they would count on seeing the brightly coloured flowers themselves whenever they were ready to turn to a new plant? If so, it looked as though it might be of interest to study their behaviour towards species with large and conspicuous flowers.

Manning, therefore, did similar tests with bees that were foraging on Foxgloves. The flowers of this species are much bigger and are also united in large inflorescences, visible from much greater distances than the Houndstongue flowers. He found that these Foxglove-bumblebees took far less trouble to learn the sites of individual plants. Nor did they react to Foxgloves without flowers or to plants of other species which had a similar general shape. When Foxglove-bees discovered a new plant, they flew straight towards it from distances up to 4 yards.

These results showed that, while bumblebees are capa

Fig. 27. Foxglove.

ble of impressive feats of learning, they applied this capacity only where it was useful. Such examples of 'planned learning', as we might call it for lack of a better term, are now known of several animals, but each new instance still astonishes one, for how could such animals know when learning pays and when it is not necessary?

Brief though this sketch of Manning's results is, yet it may show what fascinating things can still be discovered by patient observation of our commonest insects. This whole field of the relationships between plants and their pollinators has the double charm of bringing the student in contact not only with the animals, but also with the plants; only by a study of both does one appreciate the mutual relationships between them.

CHAPTER SIXTEEN

CURIOUS NATURALISTS

It is, I think, only natural for a man to have occasional doubts about the value of what he is doing; at any rate, such doubts have often occurred to me. I find studying the behaviour of animals in their natural surroundings a fascinating hobby. It allows one to live out of doors and in beautiful scenery; it gives free scope to one's urge to observe and to reflect; and it leads to discoveries. Even the most trivial discovery gives intense delight. Yet once in a while the embarrassing question comes up: 'So what?' A little devil seems to look in over one's shoulder and to take great pleasure in kindling this little spark of doubt. Reflecting about this, one tries to weigh *pro* and *contra*—and one usually ends by making up one's mind and concluding that it has all been worth while. Confidence and peace of mind are restored—but only for a while, for the doubts keep recurring.

When they do, I usually beat my little devil into retreat with the following arguments. It seems to me that no man need be ashamed of being curious about nature. It could even be argued that this is what he got his brains for and that no greater insult to nature and to oneself is possible than to be indifferent to nature. There are occupations of decidedly lesser standing.

I feel the principal gain I have had from my studies is the growing awareness of the huge variety of animal

life, the endless diversity of life patterns, of ways of cop-
ing with an adverse world; all helping life to come out of
the battle victorious. It is strange that one's insight in
this grows so slowly and also that the full significance of
observations so often dawns upon one long after they
were done. When I saw how the female Phalarope lured
her mate to the scrape where she was going to lay her
first egg, I realized that here was a nicely adapted, useful, even indispensable, behaviour pattern. But the full
awareness of this came much later, when I observed how
male Sticklebacks attain a very similar end in (naturally)
quite a different way; and when I read, and then saw
myself, how Kestrels, Redstarts, Wrens, Pied Flycatchers and the small Gobies of our tide-pools have all
evolved displays in their own way, each of which is
clearly adapted in minute detail to this vital function
of showing the nest site to their partner.

The male Kestrel sails down to the nest hole of its
choice in a beautiful glide and hooks itself on to the
rim of the hole. When you see how it spreads its tail and
when the brightness of the slate blue tail feathers with
their conspicuous black band strikes you, it suddenly occurs to you that this is probably what it has got its tail
colours for, and a similar idea occurs when you see how
the red tail of the Redstart flashes up just when it alights
in front of its nest hole. (That the female Redstart has
also got a red tail only makes you wonder whether you
have still overlooked another function.)

Or you marvel at the way a Peppered Moth blends
with the lichens on the bark of the tree trunk it has
selected; or at the way a displaying Peacock butterfly
turns itself exactly towards its enemy. Or, to take another
example almost at random, Manning's discovery that
bumblebees make a locality study when they leave a
newly-found Houndstongue plant, but do not bother to
remember the site of a Foxglove (relying so to speak on

finding it again when they want to), gave one the same kind of awe at the way these animals manage. And it was a great moment for me when Mrs Cullen began to explain how many of the peculiarities in the behaviour and the color patterns of the Kittiwake could be understood as by-products of its main adaptation—nesting on steep cliffs, inaccessible to predators.

The reward of becoming aware of the variety of life patterns comes only when one is prepared to combine concentration on one special thing with what we could call an 'open interest'. Scientific examination naturally requires concentration, a narrowing of interest, and the knowledge we gained through this has meant a great deal to us. But it has become increasingly clear to me how equally valuable have been the long periods of relaxed, unspecified, uncommitted interest. I have missed a great deal in the lives of the Bee-Wolves by being too intent on analysing their homing; and when one compares the *Philanthus* story with that of *Ammophila* it is clear that the 'open interest' of the Baerendses has paid tenfold. Also an extremely valuable store of factual knowledge is picked up by a young naturalist during his seemingly aimless wanderings in the fields.

Nor are the preliminary, unplanned observations one does while relaxed and uncommitted without value to the strict experimental analysis. I believe strongly in the importance of natural or unplanned experiments. For instance, when one sees again and again how a Black-headed Gull retreats immediately after another gull has adopted one of its threat postures, then this has almost the value of a good experiment. The difference between this and an experiment in which one uses mounted birds differing in the posture only (and not in some other respect—such as calling) is one of degree mainly and this difference is often slight. In the planned experiment all one does is to keep a few more possible variables constant; but it is amazing how far one can get by critically selecting the circumstances of one's natural experiment. By this I do not mean disregarding those that give a result contrary to expectation, but those involving too many uncertainties, just as one must disregard a laboratory experiment that happened to be disturbed by a power cut or by someone entering the room.

In the study of the ways in which animals manage to do the things that are so obviously useful to them, all *failures* are of great interest. As I mentioned in Chapter Eleven, one particular female Kittiwake invariably alighted near singing males, but always adopted the extreme form of a posture which, according to our tentative conclusions, expressed fear. She invariably fled shortly afterwards and, in fact, never acquired a mate at all. The failure of this misfit, and the correlation between the alleged fear posture and repeated overt flight, strengthened us in our conclusion that some of the postures shown even in the pair formation ceremonies of successful birds did express fear; and that fear is a tendency which normally interferes with mating and has to be overcome on penalty of failure.

I have referred to the young Herring Gull we saw systematically trying to rob Eider ducks of their eggs. It walked over an area where many incomplete clutches were lying unguarded, though well concealed under the plants the ducks put on them when they left. This gull had obviously learned to uncover eggs by clearing the cover away, but we were astonished to see that it had only the haziest idea of what a nest looked like; it wandered all over the place, moving loose plants as it went, but uncovering clutches only by accident. The other limitation in the learning capacities of gulls—showing itself in indiscriminate dropping of shell fish over soft as well as hard soil—was equally interesting. Such observations suggest that it is worth while to explore systematically not only what an animal can learn, but also what it cannot or, at any rate, does not.

'Mistaken' responses to outside stimuli are often a great help in disentangling the significant sign stimuli which release certain responses. A girl of ten once wrote to me, after I had told in a broadcast how gull chicks peck at anything red when they are hungry, that she and her little sister, walking along the shore, had encountered a full-grown young Herring Gull. It was quite tame and walked up to them—and suddenly gave a vicious peck at a 'very red scab' on the knee of the younger girl. She wrote about it because she realized that this was a kind of experiment. Similarly, an observer in Holland once reported that a very young roedeer had suddenly appeared on the road he was riding his bicycle on and persistently followed him. His bicycle had a white mudguard and I have little doubt that this kitten just obeyed its tendency to follow 'anything white'—which in nature makes it follow the white tail-patch of its mother. Still another example of the same kind of 'mistake' was observed by all the students taking a practical in animal behaviour in the lab. in Leiden: all the Stickleback males in a row of tanks dashed to the window as far as their tanks allowed them and 'attacked' a red Royal Mail van as it passed a hundred yards from the lab. responding to it as they would to a rival male, which is itself red.

Systematic exploitation of such natural experiments— that is, systematic comparison of the situations which do and those which do not release a given response—can be almost as good as planned experiments; the important thing seems to me is not to miss the natural experiments and yet to know when it becomes necessary to continue by planned tests. People who miss the tendency to watch in nature and who believe in strict experimentation, have argued that natural experiments take too long to wait for, but this is not a valid objection. Firstly, they can often be observed while one is doing the preliminary descriptive work anyway and it is better to record them than just to waste them. Secondly, if one omits the stage of observation and natural experiments, one can waste an enormous amount of time experimenting without a proper lead such as the natural experiments can provide. This may all sound very simple and straightforward (and it is), yet common sense is not always obvious.

These arguments, with which I try to silence my little devil, are admittedly simple and perhaps my devil is exceptionally naive. But that does not mean that he is necessarily mistaken. Sometimes he comes back with a counter-argument and says that, while he is tolerant enough to allow us our simple pleasures, we should not think that science will need our services very much longer. Admittedly a return to nature has been useful for a certain extension of the range of problems recognized (he says), but science has now progressed beyond this stage. My reply to this is that the usefulness of naturalistic study has not yet outlived itself and perhaps never will; I rather believe that, at every stage of biology, contact with reality will be necessary and that continued observation will time and again make us aware of new things to be explained. At every stage of research, the biologist has to be aware of the fact that he is studying, and temporarily isolating for the purpose of analysis, adaptive systems with very special functions—and not mere bits. On the other hand, I believe that many scientists are unfairly accused of failing to realize this, although admittedly some get stuck in ruts—this being a reason why so many people deny that scientific research can really enrich one's outlook on life. The curious naturalist often feels sorry for those of his fellow-men who miss such an experience; and miss it so unnecessarily, because it is there, to be seen, all the time. Nor is reading about it anything more than a poor substitute; direct, active observation is the only real thing. If my stories of curious naturalists do not send out some readers to go and observe for themselves, this book will have missed its point.

REFERENCES

This is not a full bibliography; only those publications are mentioned on which the main story is based. Most of these publications have extensive bibliographies, to which the interested reader is referred.

Chapter One
Tinbergen, N., 1932: 'Über die Orientierung des Bienenwolfes (*Philanthus triangulum* Fabr.)'. *Zs. vergl. Physiol.*, 16, 305–35.

Chapter Two
Tinbergen, N., 1934: 'Eskimoland'. Rotterdam.
Tinbergen, N., 1942: 'An Objectivistic Study of the Innate Behaviour of Animals'. *Biblioth. biotheor.*, 1, 39–98.
Schenkel, R., 1947: 'Ausdrucks-Studien an Wölfen'. *Behaviour*, 1, 81–130.

Chapter Three
Tinbergen, N., 1939: 'The Behaviour of the Snow Bunting in Spring'. Trans. Linn. Soc. N.Y., 5, 1–92.
Tinbergen, N., 1935: 'The Behaviour of the Red-necked Phalarope in Spring'. *Ardea*, 24, 1–42.

Chapter Four

Tinbergen, N. & W. Kruyt, 1938: 'Über die Orientierung des Bienenwolfes III'. *Zs. vergl. Physiol.*, 25, 292–334.

Tinbergen, N. & R. J. van der Linde, 1938: 'Über die Orientierung des Bienenwolfes IV'. *Biol. Zentralblatt*, 58, 425–35.

Beusekom, G. J. van, 1948: 'Some Experiments on the Optical Orientation in *Philanthus triangulum* Fabr'. *Behaviour*, 1, 195–225.

Tinbergen, N., 1935: 'Über die Orientierung des Bienenwolfes II'. *Zs. vergl. Physiol.*, 21, 699–716.

Chapter Five

Schuyl, G. & L. & N. Tinbergen, 1936: 'Ethologische Beobachtungen am Baumfalken, *Falco s. subbuteo* L'. *Jour. f. Ornithol.*, 84, 387–434.

Chapter Six

Baerends, G. P., 1941: 'Fortpflanzungsverhalten und Orientierung der Grabwespe *Ammophila campestris* Jur.' *Tijdschr. Entomol.*, 84, 68–275.

Adriaanse, A., 1947: '*Ammophila campestris* Latr. und *Ammophila adriaansei* Wilcke'. *Behaviour*, 1, 1–35.

Chapter Seven

Tinbergen, N., 1953: 'Social Behaviour in Animals'. London.

Tinbergen, N., 1953: 'The Herring Gull's World'. London.

Cott, H., 1940: 'Adaptive Coloration in Animals'. London (1957, 2nd Ed.).

Ruiter, L. de, 1952: 'Some Experiments on the Camouflage of Stick Caterpillars'. *Behaviour*, 4, 222–32.

Ruiter, L. de, 1956: 'Countershading in Caterpillars. An analysis of its Adaptive Significance'. *Arch. néerl. Zool.*, 11, 285–342.

Goodwin, D., 1951: 'Some Aspects of the Behaviour of

the Jay *Garrulus glandarius*'. *Ibis*, 93, 414–442; 602–25.

Kettlewell, H. B. D., 1956: 'Further Selection Experiments on Industrial Melanism in the Lepidoptera'. *Heredity*, 10, 278–301.

Ford, E. B., 1956: 'Moths'. London.

Chapter Eight

Cain, A. & P. M. Sheppard, 1950: 'Selection in the Polymorphic Land Snail *Cepaea nemoralis*'. *Heredity*, 4, 275–94.

Blest, A. D., 1957: 'The Function of Eyespot Patterns in the Lepidoptera'. *Behaviour*, 11, 209–56.

Windecker, W., 1939: '*Euchelia jacobeae* L. und das Schutztrachtenproblem'. *Zs. Morphol. Oekol. Tiere*, 35, 84–138.

Mostler, G., 1935: 'Beobachtungen zur Frage der Wespenmimikry'. *Zs. Morphol. Oekol Tiere*, 29, 381–455.

Poulton, E. B., 1890: 'The Colours of Animals'. London.

Chapter Nine

Tinbergen, N., B. Meeuse, L. Boerema & W. Varossieau, 1942: 'Die Balz des Samtfalters, *Eumenis semele* (L.)'. *Zs. Tierpsychol.*, 5, 182–226.

Ilse, D., 1929: 'Über den Farbensinn der Tagfalter'. *Zs. vergl. Physiol.*, 8, 658–92.

Chapter Ten

Besemer, A. F. H. & B. J. D. Meeuse, 1938: 'Rouwmantels'. *De Lev. Natuur*, 43, 1–12.

Chapter Eleven

Cullen, E., 1957: 'Adaptations in the Kittiwake to Cliff-Nesting'. *Ibis*, 99, 275–303.

Chapter Twelve

Moynihan, M., 1955: 'Some Aspects of Reproductive Behaviour in the Black-headed Gull (*Larus r. ridibundus* L.) and Related Species'. *Behaviour*, Suppl. 4, 1–201.

Tinbergen, N. & M. Moynihan, 1952: 'Head Flagging in the Black-headed Gull; its Function and Origin'. *Brit. Birds*, 45, 19–22.

Chapter Thirteen

Weidmann, U., 1956: 'Observations and Experiments on Egg-Laying in the Black-headed Gull'. *Brit. Jour. Anim. Behav.*, 4, 150–62.

Moynihan, M., 1953: 'Some Displacement Activities of the Black-headed Gull'. *Behaviour*, 5, 58–80.

Tinbergen, N. & A. C. Perdeck, 1950: 'On the Stimulus Situation Releasing the Begging Response in the Newly Hatched Herring Gull Chick'. *Behaviour*, 3, 1–38.

Chapter Fifteen

Manning, A., 1956: 'The Effect of Honey-Guides'. *Behaviour*, 9, 114–40.

Manning, A., 1956: 'Some Aspects of the Foraging Behaviour of Bumble-Bees'. *Behaviour*, 9, 164–203.

INDEX

ANCHOR BOOKS